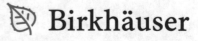

Foundations for Undergraduate Research in Mathematics

Series Editor

Aaron Wootton, Department of Mathematics, University of Portland, Portland, USA

Eli E. Goldwyn • Sandy Ganzell •
Aaron Wootton

Editors

Mathematics Research for the Beginning Student, Volume 2

Accessible Projects for Students After Calculus

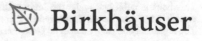 Birkhäuser

Editors
Eli E. Goldwyn
Mathematics
University of Portland
Portland, OR, USA

Sandy Ganzell
Mathematics & Computer Science
St. Mary's College of Maryland
St. Mary's City, MD, USA

Aaron Wootton
Mathematics
University of Portland
Portland, OR, USA

ISSN 2520-1212 ISSN 2520-1220 (electronic)
Foundations for Undergraduate Research in Mathematics
ISBN 978-3-031-08566-6 ISBN 978-3-031-08564-2 (eBook)
https://doi.org/10.1007/978-3-031-08564-2

Mathematics Subject Classification: 00A08, 00B10, 00-02

This book is published under the imprint Birkhäuser, www.birkhauser-science.com by the registered company Springer Nature Switzerland AG
The registered company address is: Gewerbestrasse 11, 6330 Cham, Switzerland

Preface

It is our strong belief that opportunities for mathematical research should be available to all students with an interest in mathematics—not just to those who already have expertise in the field. In our experience, engaging with research has led students to discover a talent and passion for mathematics they didn't even know they had. In fact, studies show that students who participate in early-career STEM research are more likely to remain in school, are more likely to remain in a STEM major, perform better in upper division courses, and are more interested in postgraduate STEM educational opportunities.[1] Research experience is also a desirable trait in private industry, as it illustrates an individual's abilities to problem solve.

Though there has been significant growth in research experiences in mathematics for undergraduates at 4-year colleges and research universities, these opportunities are often only available to upper-level students or those with significant background knowledge. Similar research experiences in mathematics for the beginning student—that is, those in community college and early career college students— are much rarer, even though this group stands to benefit significantly from such opportunities.

Perhaps the biggest barrier faced by the beginning student and their faculty mentors is not knowing where to begin. Accessible topics for new student researchers are often hard to find, even for experienced teachers, and finding unanswered questions that are well suited to student projects is time consuming. Accordingly, the main goal of these two volumes is to expand research opportunities in mathematics for the beginning student by removing this significant hurdle. Specifically, we seek to provide community college students, early college students, and perhaps even advanced high school students everything they need to initiate research projects with or without a faculty mentor, and to foster independence in research.

Each chapter in these two volumes is a self-contained, accessible article that provides ample background material, recommendations for further reading, and, perhaps most importantly, specific projects that can be pursued immediately upon reading the chapter. What makes these volumes different from other FURM volumes

---

[1] Hewlett, James A. *Broadening participation in undergraduate research experiences (UREs): The expanding role of the community college.* CBE-Life Sciences Education 17.3 (2018): es9.

is that the chapters have been written for the beginning student by minimizing the number of prerequisites. Indeed, many of the chapters require no prerequisites other than a desire to pursue mathematics, and even the most ambitious chapters require no more than linear algebra or introduction to proofs, both of which are typically sophomore-level classes.

This volume is geared towards students with at least some exposure to calculus, with the later chapters requiring some additional prerequisites such as linear algebra and statistics.

Exploring mathematics is a lot like playing with toys—at least, it can or should be, and Chapter "Constructible Pi and Other Block-Based Adventures in Geometry" stands as direct (and hopefully clear) proof of this. In it, Douglas explores a collection of questions and games that use nothing more than building blocks to answer and play. Moreover, these projects enjoy deep connections to both historical and current mathematics, providing the engaged reader plenty of cover when a passerby asks, "Why are you sitting around playing with all of those little bricks?" You could inform them that you are trying to maximize a certain amount of enclosed space and invite them to join you! Or you could shoo them away with an equivalent, but less inviting response like this one: "I am conducting serious research into various, discrete generalizations of the isoperimetric problem."

Chapter "Numerical Simulation of Arterial Blood Flow," traverses the inter-disciplines of mathematics and medicine inspired by modeling blood flow through a blood vessel, using Poiseuille's law. In this chapter, Pati and Ladipo develop a one-dimensional numerical scheme to solve the differential equations establishing a relationship between velocity of blood and the radius of the vessel. With historical narratives, this project will inspire any calculus student to explore the field of biomedical research.

In Chapter "Statistical Tools and Techniques in Modeling Survival Data," Overman and Pal introduce the rich field of survival analysis that is essential to the fields of medicine, industrial engineering, and more. Survival analysis studies the expected times of events occurring, such as the recurrence of a tumor in a cancer patient or the time to failure of a manufactured component. The authors describe ways of modeling these phenomena to make well-informed predictions and provide code in R to apply these techniques to real data. Throughout the chapter, there are exercises of various difficulties to keep the reader engaged. The chapter closes with computing projects on given data as well as a more difficult open problem in the field.

Option pricing has garnered a lot of attention in the last five decades, especially with the advancement of the Black-Scholes-Merton approach. Large sums of money are staked every day on the price movement of stocks, commodities, and other assets, and this shows no sign of abating anytime soon. In Chapter "So You Want to Price and Invest in Options?", Cohen and Loke introduce readers to the binomial model for asset pricing, and its extension to investing via maximal expected utility and valuing insurance premiums via indifference pricing. Illustrative examples are provided, and projects for interested students that emphasize calibration and statistical analysis are also given. It is the authors' sincere hope that readers exit this chapter with the same enthusiasm they have for this beautiful area at the intersection of quantitative analysis, finance, and data science.

In Chapter "The Spiking Neuron", Bohling and Udeigwe take the reader through a survey of a class of models for the "The Spiking Neuron." The nervous system is constantly responding to stimuli as it interacts with the surrounding environment. If the intensity of a stimulus reaches a certain threshold specific to a group of neurons, these neurons are activated to transmit electrical signals. The graph of the membrane potential of these neurons, during this period of signal transmission, has a spike shape. The authors go through the derivation and computational implementation of exemplar models that describe this neuronal behavior, complete with definitions and exercises. They also provide pieces of Python code to help students with the programming exercises and suggested projects.

In Chapter "Counting Lattice Walks in the Plane," Klee introduces counting problems that are motivated by a classical combinatorial question: in how many ways can one walk from the origin to a given point in the xy-plane using steps that only go north or east by one unit? An extension of this problem asks for the number of ways to reach a given terminal point if one is only allowed to take steps that come from a prescribed set of integer vectors. This chapter guides the reader through a range of combinatorial proof techniques that have been useful in exploring these problems and highlights the work of several groups of undergraduate researchers.

In Chapter "The Mathematics of Host-Parasitoid Population Dynamics," Emerick explores both discrete and continuous mathematical models of the host-parasitoid interaction. Typically regarded as something only Hollywood could dream up, the parasitoid is a parasitic insect that ultimately kills its host by using it as a vehicle for reproduction. In areas where the host species is a pest to specific plant or crop species, parasitic wasps can be introduced as a form of natural pest control. Determining specific behaviors and characteristics that lead to the coexistence of both the parasitoid and host population is a key interest in biological pest control. We introduce the reader to the basic mathematical models associated with host-parasitoid ecology and explore different models that include a discrete and continuous component. Various computer program templates are included and a variety of different parasitoid behaviors are explored along the way. After working through the chapter, the reader will be well equipped to come up with a new model of host-parasitoid dynamics.

In Chapter "Mathematical Modeling of Weather and Climate," Pendleton explores the ways in which we can use calculus and statistics to study weather and climate change. Techniques for the development of mathematical models as they relate to Earth's climate system and weather phenomena are proposed. These ideas are illustrated and developed through real-world examples using computational experiments and data analysis.

Chapter "Beyond Trends and Patterns: *Importance of the Reproduction Number from Narratives to the Dynamics of Mathematical Models*," explores an important quantity in epidemiology and mathematical biology. This quantity is often referred to as the tipping point or reproduction number and appears in different narratives throughout medical, social, and life sciences. In this chapter, Ghosh and Mubayi present various models for the reproduction numbers in different contexts. They also discuss applications of the reproduction number in designing intervention strategies through testing and vaccination. Showing different applications of reproduction numbers will allow readers from multidisciplinary domains to understand the power and importance of this concept.

Finally, Chapter "Application of Mathematics to Risk and Insurance," lies at the intersection of insurance, probability theory, and statistics. Auto-insurance companies use individualized risk approaches to determine whether to insure an applicant and which premium to charge. The approach is known as the Bonus-Malus system (BMS). In this chapter, Das provides a brief description of three commonly used claim frequency models. He has also discussed a few BMSs used in several countries around the world. He explores the concept of the stationary distribution of a BMS with a numerical example.

Portland, OR, USA   Eli E. Goldwyn
St. Mary's City, MD, USA   Sandy Ganzell
Portland, OR, USA   Aaron Wootton

Contents

Constructible Pi and Other Block-Based Adventures in Geometry

Casey Douglas

Abstract

The Isoperimetric Problem is as classical a mathematics topic as it is revolutionary. This chapter explores an interesting discrete version of this ancient subject, one ripe with opportunity for continued work and rewarding LEGO®-based play, including new ways to conceive of and construct notions of π.

Suggested Prerequisites *High School/College Algebra, Arithmetic, Some Calculus, Some Discrete Mathematics*

1 Introduction

Have you ever wondered about the constant $\pi \approx 3.14159\ldots$? Me, too, and so have lots of other people [11]. In fact, this number has intrigued humankind for millennia, and it continues to do so. From awe inspiring formulas like

$$\pi = \frac{2}{1} \cdot \frac{2}{3} \cdot \frac{4}{3} \cdot \frac{4}{5} \cdots \quad \text{and} \quad e^{i\pi} + 1 = 0$$

to uses in art, architecture, and various other endeavors (not all beginning with "a" including [2, 7], and [10] for instance), it is doubtful that we will soon if ever tire of it or cease to find it useful. But this all begs a rather fundamental question: what

C. Douglas (✉)
Department of Mathematics, University of Houston, Houston, TX, USA
e-mail: cdouglas@uh.edu

© The Author(s), under exclusive license to Springer Nature Switzerland AG 2022
E. E. Goldwyn et al. (eds.), *Mathematics Research for the Beginning Student,*
Volume 2, Foundations for Undergraduate Research in Mathematics,
https://doi.org/10.1007/978-3-031-08564-2_1

exactly is π, anyway? Just *how* is it defined and *why* is it defined *that* way? Does it remain just as special if these choices are altered or adapted to different situations?

The primary goal of this chapter is to contemplate some answers to these questions by way of so-called Isoperimetric Problems, but to do so in ways that encourage creativity, benefit from abstraction, and require an active audience. It will come as no surprise, then, that our exploration will use LEGO® bricks, deliberately blurring the already-blurred-into-oblivion line separating *mathematics* from *play*. The readers and active participants will be treated to a pi-lethora of tools, toys, and ideas, some sophisticated and modern, others classical or quaint, but all in service of a foundation for continued growth and imaginative play, i.e., mathematical research.

We begin in Sect. 1.1 with a review of the standard (and original) definition of π as a geometric quantity that compares any circle's circumference to its diameter. This leads naturally to Sect. 1.2 with an accessible summary of the (planar) Isoperimetric Problem wherein both circles and π reappear as key players. By using building blocks to discretize this classical problem, co-explorers will discover how to define and understand LEGO®-based circles and π. Such discoveries are facilitated by the inclusion of Definitions, Exercises, and Suggested Activities; a research mentor would do well to treat the material in Sects. 1.1 and 1.2 as introductory reading they might assign interested students before the start of a collaborative research project.

Section 2.1 uses Definitions and Suggested Activities to help phrase the primary source of related research projects, namely the Block Isoperimetric Problem and the Dual Block Isoperimetric Problem. Throughout Sects. 2.2–2.4, a sequence of guided exercises is discussed in order to help participants understand key features of these problems. The entirety of Sect. 2 serves as a kind of guide or model for mentors and students alike as to how these sorts of problems can be approached and generalized. Finally, Sect. 3 collects a variety of open research problems related to the problems discussed in Sect. 2.

Note The suggested activities, exercises, and challenge problems featured in this chapter are best facilitated by a research mentor or instructor but can safely be attempted by students learning independently. The guided exercises, on the other hand, are based on the work done by two REU groups hosted at St. Mary's College of Maryland in 2010 and 2012; these exercises are designed to facilitate both raw mathematics skills and communication (i.e., proof-writing) skills. It should be noted that the eight REU students participated in these projects with backgrounds consisting of, at most, two semesters of Calculus. Indeed, this topic does not actually require a full semester of Calculus to study, although some familiarity with optimization will help as would experience in courses like Discrete Mathematics.

For top notch and friendly introductions to the Isoperimetric Problem, we strongly recommend Weigert's award winning article [13] and Blasjö's article [5]. A different discrete version of the Isopeirmetric Problem was posed and studied by Miller and Morgan in [9], while works done by Chakerian [6] and Wallen [12] contemplate other novel versions of this problem. The resources mentioned here are suitable for early career students. In addition, students and instructors interested in resources on discrete mathematics and proof-writing are encouraged to use the books [8] and [4].

Special Note to Students Hi, and thanks for reading through all of this so far. I am sure you are itching to dive right into some mathematics as quickly and meaningfully as possible, and to this end it is recommended that Sects. 1.1, 1.2, and 2.2–2.4 be lightly skimmed, outright skipped, or initially delayed, at least for a first pass-through. Jumping straight from here to the beginning of Sect. 2 and then to Sect. 3 should give you a great sense of what these problems and projects are all about.

1.1 In the Beginning There Were Circles

Most high school and college students are familiar with the unit circle as well as the famous equation $x^2 + y^2 = 1$ that gives rise to it, but for our purposes it is only necessary to recall that most essential or elementary definition of a circle as a special set of points in $\mathbb{R}^2$.

Definition 1 A **circle** is a set of points each located an equal distance from a fixed point (called the center). More precisely, a circle with radius r and center P is the set of points (in the plane) whose distance to P equals r.

Exercise 1 Use the distance formula/the Pythagorean theorem to explain how the equation $x^2 + y^2 = 1$ matches with the definition above to produce the familiar unit circle.

Suggested Activity 1 Use compasses to sketch circles of different radii. Approximate the circumference of each circle by measuring the length of a piece of string that wraps around each one. Finally, compare the ratios of circumference to diameter.

Despite our familiarity with these objects, it often comes as something of a surprise to students that these ratios are, in fact, very close. It is worth seizing upon this excitement to point out that any two circles are *similar*—one circle can always be rescaled to look exactly like another—and that this scale invariance is what is behind their identical ratios. Of course, *this* common ratio is the original definition of π.

Definition 2 The constant π is defined to be the ratio of any circle's circumference, C, to its diameter, d: $\pi = C/d$.

It is also worth noting that from this *definition*, one easily recovers the familiar formula for a circle's circumference in terms of its radius.

Exercise 2 Among the options provided below, which other quantities related to a circle are also scale invariant? (r, d, and C denote a circle's radius, diameter, and circumference, respectively, and A denotes the area it encloses.)

(a) r/d
(b) r^2/A
(c) C/r^2
(d) $C^2/(4A)$

Of these scale invariant quantities, which ones (if any) equal π? What other scale invariant quantities (not listed here) can you come up with?

Challenge Problem 1 Our definition of π relies on our definition of circles which itself relies on our standard definition of *distance*. If we change this definition of distance, how might the value and meaning of π change? In particular, if we instead use the "taxicab distance"

$$\text{dist}_{\text{TC}}\big((x, y), (u, v)\big) = |x - u| + |y - v|,$$

what do circles look like? Is the quantity C/d still scale invariant? If so, what is the value of π in this setting?

Challenge Problem 2 What do circles look like when we use the so-called chessboard notion of distance? The formula for *this* distance function is given by

$$\text{dist}_{\text{CB}}\big((x, y), (u, v)\big) = \max\big\{|x - u|, |y - v|\big\}.$$

Is the quantity C/d still scale invariant? If so, what is the value of π in this setting?

Fig. 1 LEGO® Dido

1.2   The Isoperimetric Problem and the Story of Queen Dido

Legend has it that, a long time ago in a land far, far away, a clever woman accumulated an impressive amount of land on which to establish her Queendom, and she did so using nothing more than strips of ox hide. As discussed in [5], this is the story of how Queen Dido established the city of Carthage (so its perhaps not so "far, far away" after all), and it is not as bizarre as it first sounds, either (Fig. 1).

Queen Dido's ship landed on the coast of North Africa where she bargained for land. Although the natives denied her original requests, they *did* agree to some rather peculiar terms: the once-and-future Queen would be given as much land as an ox's hide could surround. Our mathematically gifted heroine accepted this challenge, cutting the skin into thin strips and arranging them in the shape of a semicircle. Thus, not only was the great city of Carthage born, but so too was the *Isoperimetric Problem*:

The Isoperimetric Problem: what is the largest amount of area that can be enclosed by a curve of a fixed length ℓ, and which curve(s) accomplish this?

Exercise 3 Why is the neologism "isoperimetric" used to name this problem?

Exercise 4 Consider an equilateral triangle and a square each with perimeter $\ell = 1$. Which shape encloses more area?

Challenge Problem 3 (The Triangular Isoperimetric Problem) Explain why/show that among all triangles with perimeter ℓ, the equilateral triangle encloses the most amount of area $A = \ell^2\sqrt{3}/36$.

Dido did not posit a triangular-shaped piece of land (nor any other polygonal shape) when demarcating the future city of Carthage. She correctly intuited that a circular arrangement would enclose as much land as possible, and, indeed, a circle of radius $r = \ell/(2\pi)$ uniquely answers the Isoperimetric Problem. A related Dual problem is given as follows:

The Dual Isoperimetric Problem (aka The Iso-area Problem): what is the smallest possible perimeter of a curve that encloses a fixed amount of area A, and which curves achieve this?

A circle of radius $r = \sqrt{A/\pi}$ also answers this question. Rigorous proofs of these claims were centuries in the making, however, and over the years, various generalizations of them have been studied. The suggested activity below provides strong evidence as to why a circle solves the Isoperimetric Problem.

Suggested Activity 2 Using nothing more than a flat soap film and a loop of thread, one can provide compelling evidence that these Isoperimetric Problems are solved by the circle. Use a wire frame bent into the shape of a square, and dip it into soapy water. Next, place the loop of thread onto the soap film, and then puncture the interior of the loop. The soap film, desiring to minimize surface area/surface tension, will pull the looped thread into a shape that contains as *much* area as possible.

Of particular note is the fact that the Isoperimetric Problem and its Dual are both *uniquely* solved by circles, giving rise to alternate definitions of both circles and π. Given a closed curve γ, we will continue to let $\ell(\gamma)$ denote its length and let $A(\gamma)$ denote the amount of area it encloses.

Definition 3 (Isoperimetric Definition of a Circle) A closed curve γ is a **circle** if and only if among all other curves β with the same length (i.e., $\ell(\gamma) = \ell(\beta)$), it follows that β encloses less area than γ (i.e., $A(\beta) \leq A(\gamma)$).

Definition 4 (Iso-Area Definition of a Circle) A closed curve γ is a **circle** if and only if among all other curves β enclosing the same amount of area as γ (i.e., $A(\beta) = A(\gamma)$), it follows that β is longer than γ (i.e., $\ell(\beta) \geq \ell(\gamma)$).

Moreover, the Isoperimetric Problem and its solutions imply that for any closed curve γ

$$\frac{\ell(\gamma)^2}{A(\gamma)} \geq 4\pi$$

with equality if and only if γ is a circle. This observation leads naturally to the following alternate definition of π.

Definition 5 (Isoperimetric Definition of π) The constant π is the minimum value of

$$\frac{\ell(\gamma)^2}{4A(\gamma)}$$

where the minimum is taken over *all* possible closed curves γ.

It must be remarked that this is, perhaps, one of the *worst* ways to define π, especially for beginning mathematicians. A direct use or test of this definition would first require one to measure the perimeters and enclosed areas of all possible closed curves, ever! The important takeaway is not that one *should* always define circles or π in these ways, but that because of the unique geometric properties these objects possess one *could* do so. In fact and as we shall soon see, the discrete block curves developed in Sect. 2 are well suited to such definitions.

Exercise 5 Let γ be a curve that encloses $A(\gamma)$ units of area and that measures $\ell(\gamma)$ units of length. Explain why/show that the quantity $\ell^2(\gamma)/A(\gamma)$ is scale invariant.

Challenge Problem 4 Use the fact that circles uniquely solve the Isoperimetric Problem to explain why/show that for any closed curve γ the so-called Isoperimetric Inequality

$$\frac{\ell(\gamma)^2}{A(\gamma)} \geq 4\pi$$

holds with equality if and only if γ is itself a circle.

Challenge Problem 5 Compute the value of π using Definition 5 and using the taxi-cab distance mentioned in Challenge Problem 1. (A slightly generalized version of this question was studied in [12].)

2 Block Headed Thinking

Imagine being given a bucket of 1×1 LEGO® tiles and a large base plate on which to arrange them. How might a sequence of these tiles be arranged so as to enclose as

many unused spaces as possible? If you are thinking that this question has the feel
of a suggested activity, I agree, but it comes with some illuminating caveats.

Suggested Activity 3 Take a set amount of 1×1 tiles (or $1 \times 1 \times 1$ bricks)
and empty base plates. It is probably best to first use 20 bricks and then use
21 bricks, for instance, or, if multiple participants are involved, to have some
use 20 and others use 21. Under a given time constraint, attempt to enclose
the largest amount of space possible using arrangements of the given bricks.

Careful readers may quickly note that the terms "arrangement" and "enclose"
need clarification, and while this is certainly true, it is also true that most people
will infer rather standard interpretations of these terms. Roughly, the blocks will
need to form a contiguous sequence and the number of unused spaces will be used
as the amount of "enclosed area."

Participants should explain how they attempted to arrange their bricks and what
their results were. It is probable (although by no means certain) that some will have
achieved the maximum possible area, namely 16 bricks, and they are likely to have
done so using a 5×4 rectangular arrangement. But why should they believe that
this is the maximum possible area?

The participants assigned 21 bricks will have noticed that, try as they might, only
at most 20 of their bricks were useful when attempting to enclose a large amount
of area. Before reading on, it is worth contemplating how one's strategies would
change if instead of 20 or 21 blocks they were given 100 or 101, or even 5, 550 or
5, 551 blocks to use. Does an even or odd number of blocks make a difference? How
does the number of block's parity play into this?

A related or dual activity is to reverse the previous setup and instead assign a
fixed amount of area to enclose using the *fewest* number of bricks.

Suggested Activity 4 Given a very large collection of 1×1 tiles (or $1 \times 1 \times 1$
bricks) and empty base plates, attempt to enclose 8 units of area and then
9 units of area using the fewest number of bricks. Record your results and
strategies.

Both activities, of course, speak to the central topic of this chapter, a LEGO®
Isoperimetric Problem, precise details for which we present in the following
subsection.

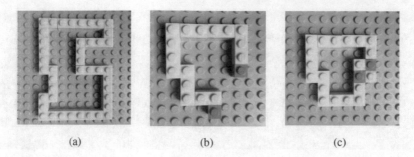

(a) (b) (c)

Fig. 2 Examples of block curves and non-block curves. (**a**) Block curve. (**b,c**) Non- block curve

2.1 Block Curves and Block Isoperimetry

To formalize matters, we offer the following definitions and emphasize that visualizing or constructing examples are likely the best way to gain a solid understanding.

Definition 6 (Blocks and Neighbors) A **block** is a unit square (whose sides can be taken to be parallel to the x- and y-axes). Moreover, we say that two blocks are **neighbors** if they share a side.

In easier-to-visualize LEGO® terms, we can think of a block as a 1×1 tile (or a $1 \times 1 \times 1$ brick), and two such blocks are neighbors if they occupy adjacent spaces on a base plate.

Definition 7 (Block Curves) A **block curve** is a finite collection of blocks, β, wherein each member of the collection has exactly two neighbors.

A top-down view of some block curves and some non-block curves is depicted in Fig. 2. The arrangement depicted in Fig. 2a can and should be verified to satisfy the key property that every block contains exactly two neighbors.

The arrangements depicted in Fig. 2b,c are not block curves since they contain blocks (shown in a different color/shade) that do not have two neighbors (the former featuring blocks with only one neighbor and the latter having blocks that contain 3 neighbors).

Naturally, we will refer to the number of blocks contained in a block curve β as its *length* or *perimeter*, and we will refer to the number of unused LEGO® spaces surrounded by β as its *enclosed area*. For instance, Fig. 2d shows a block curve β with perimeter $P(\beta) = P = 42$ and area $A(\beta) = A = 38$. Whenever possible, we will elect to denote a block curve's perimeter and area by P and A, omitting their explicit dependence on β.

Suggested Activity 5 Using paper and pen or, even better, LEGO® bricks, create several examples of block curves. You are likely to observe that their perimeters are always *even*. Does this have to happen, or is it ever possible for *P* to be odd? If an odd perimeter is possible, provide an explicit example. If you are convinced only even perimeters are allowed, think about designing a step-by-step algorithm or computer code that verifies this hypothesis.

With notions of perimeter, area, and block curves under our belts, we are almost ready to pose our block Isoperimetric Problems. First, though, we should set up some useful notation via exercise 6:

Exercise 6 The **floor** of a number x is denoted and defined by

$$\lfloor x \rfloor = \text{the largest integer less than or equal to } x.$$

Similarly, the **ceiling** of a number x is denoted and defined by

$$\lceil x \rceil = \text{the smallest integer greater than or equal to } x.$$

(a) Evaluate the expressions $\lceil 2 \rceil$, $\lfloor -5 \rfloor$, $\lfloor 1/2 \rfloor$, $\lceil \pi \rceil$, $\lfloor e \rfloor$, and $\lfloor \pi \rfloor - \lceil e \rceil$.
(b) Plot graphs of the functions $y = \lfloor x \rfloor$ and $y = \lceil x \rceil$.

Without further ado, ladies and gentlemen, the stars of our show:

The Block Isoperimetric Problem. What is the largest amount of area a block curve of fixed perimeter P can enclose, and which block curves achieve this maximum area?

The solution to the first part of this problem is as follows: given P perimeter blocks, the maximum area one can enclose is given by

$$\text{Max}(P) = \left\lceil \frac{P}{4} - 1 \right\rceil \cdot \left\lfloor \frac{P}{4} - 1 \right\rfloor. \tag{1}$$

The Dual Block Isoperimetric Problem (aka The Iso-area Problem). What is the smallest possible perimeter of a block curve that encloses a fixed amount of area A, and which block curves achieve this minimum perimeter?

The solution to the first part of this problem is as follows: the smallest number of perimeter blocks needed to enclose A units of area is given by

$$\text{Min}(A) = 2 \left\lceil 2\sqrt{A} \right\rceil + 4. \tag{2}$$

While it may seem that both questions are merely two sides of the same coin, this is not the case, at least not in the sense that both questions are equivalent. Fixing perimeter and maximizing area *or* fixing area and minimizing perimeter are equivalent problems in the classical setup, both leading to Dido's circle, but, interestingly enough, the answers to The Block Isoperimetric Problem and its Dual do not always agree.

As an example of this phenomenon, consider the following exercise.

Exercise 7 Use the formulas for $\text{Max}(P)$ and $\text{Min}(A)$ provided in Eqs. (1) and (2) above to answer the following questions:

(a) Determine the largest amount of area A that can be enclosed using $P = 26$ bricks.
(b) Determine the smallest amount of perimeter P needed to enclose $A = 30$ units of area.

This phenomenon warrants further study, presented here as a challenge problem.

Challenge Problem 6 Determine those values of P and A for which

$$\text{Min}\Big(\text{Max}(P)\Big) = P$$

$$\text{Max}\Big(\text{Min}(A)\Big) = A.$$

Back to formulas (1) and (2), which are not supposed to be obvious, and so apologies are in order for having them thrust upon unsuspecting readers. To this point, the bulks of Sects. 2.2 and 2.3 contain a list of guided exercises designed to demonstrate precisely where these formulas come from and the tools or approaches used to discover them.

Before entering those discussions, though, we collect a handful of final definitions, challenge problems, and properties of block curves and circles.

Definition 8 (Block Circle) A block curve with P units of perimeter that encloses A units of area is a **block circle** if and only if one of the following conditions holds:

1. $A = \text{Max}(P)$ as given in Eq. (1)
2. $P = \text{Min}(A)$ as given in Eq. (2).

Challenge Problem 7 Find two different block circles with perimeters $P = 16$ and enclosed areas $A = 8$. Here, "different" means that neither of the two block curves can be translated, rotated, or reflected to coincide with the other.

Definition 9 ($\boxed{\pi}$) The constant $\boxed{\pi}$—pronounced "**Block** π"—is the minimum value of the ratio

$$\frac{P(\beta)^2}{4A(\beta)},$$

where the minimum is taken over all possible block curves β.

Exercise 8 Show that the ratio $P(\beta)^2/A(\beta)$ can change even when β is a block circle. This can be accomplished by finding two different block circles with different perimeter/area combinations and comparing the different values of this ratio.

Challenge Problem 8 Use formulas (1) and (2) for $\text{Min}(P)$ and $\text{Max}(A)$ (along with any necessary Calculus) to conclude that

$$\lim_{P \to \infty} \frac{P^2}{\text{Max}(P)} = \lim_{A \to \infty} \frac{\text{Min}(A)^2}{A} = 16.$$

Use these limits to conclude that $\boxed{\pi} = 4$. Compare this value of $\boxed{\pi}$ with the values of π from Challenge Problems 1, 2, and 5.

Definitions 8 and 9 are strikingly similar to/transparently inspired by Definitions 3, 4, and 5. As promised, we conclude this section with a list of useful and interesting or surprising facts about block circles and $\boxed{\pi}$. References to specific guided exercises, challenge problems, and sections in the table below indicate that the property can be explained by carefully completing those tasks.

Block curve properties	
$\text{Max}(P) = \lceil P/4 - 1 \rceil \cdot \lfloor P/4 - 1 \rfloor$	Guided Exercises 1–3
$\text{Min}(A) = 2 \lfloor 2\sqrt{A} \rfloor + 4$	Guided Exercises 4 and 5
Block circles are not unique	Chall. Prob. 7 and Sect. 2.4
Some (but not all) block circles are rectangular	Guided Exercise 6–8
P^2/A is not constant on block circles	Exercise 8
$\lfloor \pi \rfloor = 4$	Chall. Prob. 8

2.2 The Block Isoperimetric Problem

This section is devoted to deriving formula (1), which we restate here as a theorem.

Theorem 1 *Among all block curves with perimeter P, the maximum amount of area that can be enclosed is*

$$A = \left\lceil \frac{P}{4} - 1 \right\rceil \cdot \left\lfloor \frac{P}{4} - 1 \right\rfloor.$$

Moreover, a rectangular block curve obtains this maximum.

Whereas the term "theorem" refers to a mathematical fact that comes equipped with an explanation or "proof," the term "lemma" is often used as the name for a kind of "miniature theorem" that is primarily used to help explain a larger, full-blown theorem. To understand why Theorem 1 is true, we will proceed by first establishing two bite-sized lemmas, each of which we present as to-be-completed guided exercises.

Guided Exercise 1 Every block curve has an even perimeter

Step 1. For an arbitrary block curve β with perimeter P, label its individual blocks as $b_1, b_2, \ldots, b_P$ in such a way so that b_2 is a neighbor of b_1, b_3 is a neighbor of b_2, and, in general, b_{i+1} is a neighbor of b_i.
Step 2. Explain why it follows that b_P is a neighbor of b_1.
Step 3. One can think of "moving" from b_1 to b_2 by sliding the tile b_1 into the location occupied by b_2 and similarly b_i into b_{i+1}'s position.
Step 4. Explain why the total number of horizontal and vertical slides needed to move b_1 along β and ultimately into b_P's position both is necessarily even.
Step 5. Conclude that the total number of blocks must be even.

Guided Exercise 2 Given any block curve β with perimeter P, there exists a rectangular block curve of perimeter P that encloses as much area as β, if not more

Step 1. Define and record the largest horizontal distance separating any two blocks in β (this distance can be achieved by multiple pairs of blocks) and label this quantity w.

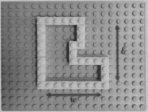

Step 2. Deconstruct the block curve β and use its blocks to build a new, rectangular curve β' that encloses a rectangular region of width w and length ℓ.

Step 3. By construction, β' will enclose more area (compare Fig. 3a,b), and, as indicated in Fig. 3b, there may be left over blocks. However, by Guided Exercise 1, we know that the perimeters of both β and β' are even and so there are always an *even* number of leftover blocks from Step 3.

Step 4. Our goal is to maximize the area, but we must also maintain perimeter. To accomplish this, use the even number of left over blocks to extend β' into a rectangular block curve that possibly encloses even *more* area. This process is depicted in Fig. 3b through Fig. 3f.

Guided Exercise 3 Proving Theorem 1

Step 1. Guided Exercise 2 allows us to derive the claimed formula for maximum area under the added assumption that our block curve β is a rectangular curve.

Step 2. Explain why the rectangular block curve β must enclose a rectangular region consisting of x number of columns and y number of rows as well as 4 corner blocks where $P = 2x + 2y + 4$.

Step 3. Our task is to maximize the enclosed area $A = xy$ subject to the constraint $P = 2x + 2y + 4$. Explain why this is equivalent to maximizing

(continued)

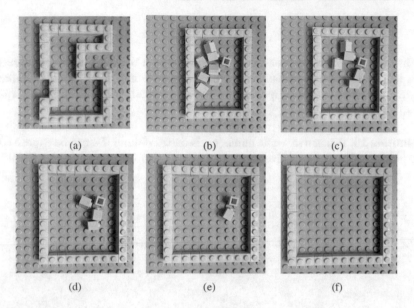

Fig. 3 Increasing area using rectangular curves. (**a**) Initial curve β. (**b**) Rectangular curve β'. (**c**) Extended β'. (**d**) Extended β'. (**e**) Extended β'. (**f**) Extended β'

the function

$$A(x) = x\left(\frac{P}{2} - 2 - x\right).$$

Step 4. The maximum value of $A(x)$ occurs when $x = P/4 - 1$, which is not necessarily a whole number. To correct for this, we can set x equal to either the ceiling or the floor of this expression, but whatever choice is made for x, show that the remaining option is found to be the value of y.

Step 5. Conclude that the resulting maximal area is then given by the desired formula (1).

2.3   The Dual Block Isoperimetric Problem

This section is devoted to deriving formula (2), which we also restate here as a theorem.

Theorem 2 *The smallest possible perimeter of all block curves enclosing a given area A is*

$$P = 2\lceil 2\sqrt{A}\rceil + 4.$$

In order to guide interested co-explorers through the derivation of this result, it is helpful to use the results from the previous section to determine the minimum perimeter needed to enclose *special* kinds of area, namely when A is a perfect square or an *oblong number*.

Definition 10 A positive whole number a is called **oblong** if it can be expressed as the product of two consecutive numbers, i.e., $a = n(n + 1)$ for some $n \in \mathbb{N}$.

Guided Exercise 4 The smallest possible perimeter of all block curves enclosing a square or oblong area $A = xy$ is given by

$$P = 2\lceil 2\sqrt{A}\rceil + 4 = 2x + 2y + 4.$$

Step 1. When $A = xy$ is a square, this means $x - y$ and the claimed formula

$$2\lceil 2\sqrt{A}\rceil + 4 = 2x + 2y + 4$$

is straightforward to verify. However, when $A = xy$ is oblong and, say, $x + 1 = y$, one should verify that the claimed equation is equivalent to checking

$$2x < 2\sqrt{x^2 + x} \leq 2x + 1.$$

Step 2. It is worth noting that the proposed minimal perimeter can be obtained by arranging the area $A = xy$ as a square or as an oblong rectangle. Guided Exercise 2 from the previous section allows us to conclude that these curves enclose as much area as possible.

Step 3. Explain why the expression for maximum area subject to perimeter P from Eq. (1) is strictly increasing in the variable $P \in \{8, 10, 12, \dots\}$. Indeed, this behavior is depicted in Fig. 4.

Step 4. It now follows that if $P' < P$, then the maximum amount of area enclosed by a curve with perimeter P' is strictly less than the maximum amount of area enclosed by a curve with perimeter P. As a result, only curves with perimeters given by the proposed formula can enclose the stipulated $A = xy$ units of area.

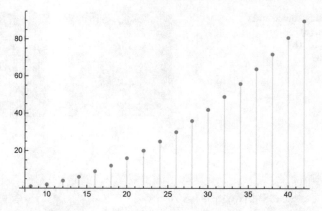

Fig. 4 Plot of $\mathrm{Max}(P)$ for even P

Guided Exercise 5 below establishes the veracity of Theorem 2, and it does so in ways that also use *square* and *oblong* numbers. In preparation for these, we offer the following definitions, and these concepts will enjoy more use in Sect. 2.4.

Definition 11 (Nearest Squares and Oblong Numbers) Given a positive number A, denote by A^* and A_* the following quantities:

$A^* =$ the smallest square or oblong number that is greater than or equal to A.

$A_* =$ the largest square or oblong number that is smaller than or equal to A.

Exercise 9 Evaluate the quantities 5^*, 5_*, 6^*, 6_*, 24^*, and 24_*.

Challenge Problem 9 Verify the following formulas:

$$A^* = \left\lceil \frac{A}{\left\lceil \sqrt{A} \right\rceil} \right\rceil \cdot \left\lceil \sqrt{A} \right\rceil$$

$$A_* = \left\lceil \frac{A}{\left\lceil \sqrt{A} \right\rceil} \right\rceil \cdot \left\lceil \sqrt{A} - 1 \right\rceil .$$

Fig. 5 Indenting square and
oblong curves. (**a**)
$A^* = 9$, $P = 16$. (**b**)
$A = 8$, $P = 16$

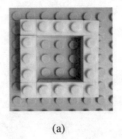

(a)   (b)

Guided Exercise 5 Proving Theorem 2

Step 1. Given a desired area, A, apply Guided Exercise 4 to the new area A^*.
Step 2. Rearrange or indent the square or oblong block curve enclosing A^*
       units of area to produce a new curve enclosing A units of area. (This
       process is depicted in Fig. 5 with $A = 8$ and $A^* = 9$.)
Step 3. The monotonicity of $\text{Max}(P)$ ensures that all smaller perimeters
       enclose strictly less area, allowing us to conclude that $\text{Min}(A^*) = \text{Min}(A)$
       and thereby verifying the desired formula for minimum perimeter.

2.4   Isoperimetric Pairs

In this section we address a phenomenon that one is likely to have noticed after
much playing and building and thinking upon the Block Isoperimetric Problem (and
its Dual).

As we have already noted, the best way to enclose 8 units of area is to use 16
blocks, but now a bit more can be said. In accordance with our walk-through of
Theorem 2, we are able to enclose 8 units of area by constructing an "indented
square" whose perimeter is 16. However, this is not the *only* way to minimize
the perimeter. As depicted in Fig. 6, a rectangular arrangement of 16 blocks *also*
encloses 8 units of area.

When does this happen in general? That is, *which rectangular arrangements
of area are enclosed by a minimum-perimeter block curve, i.e., by a block circle?*
According to Guided Exercise 4, all perfect squares and oblong numbers will have
this property, but as our consideration of $A = 8$ demonstrates there are other
numbers with this property, too. There are lots of them, in fact, and we will give
them a name:

Definition 12 (Block Number) A natural number A is said to be a **Block Number**
if there is a rectangular region containing A units of area that is enclosed by a block
circle.

Fig. 6 Rectangular curve
with $A = 8$

The first 23 block numbers are

$$1, 2, 3, 4, 6, 8, 9, 10, 12, 15, 16, 18, 20, 21, 24, 25, 28, 30, 32, 35, 36, 40, 42, \ldots$$

and no (currently listed) sequence in the Online Encyclopedia of Integer Sequences (OEIS) contains this list. At first glance, the sequence of block numbers appears to be devoid of much structure, but quite a bit can be demonstrated about these numbers. In particular, one can show the following:

Theorem 3 *A number A is a block number if and only if $A^* - A$ is zero or is of the same "type" as A^* (i.e., square or oblong).*

For instance, the number 15 is a block number since $15^* = 16$ is a perfect square and $16 - 15 = 1$ is also a perfect square. Similarly, 50 is a block number since both $50^* = 56$ and $56 - 50 = 6$ are oblong.

There are two key ingredients needed to prove Theorem 3, both of which we present as guided exercises.

Guided Exercise 6 **A number A is a block number if and only if there exists a pair of divisors $n \cdot m = A$ satisfying**

$$m + n = \left\lceil 2\sqrt{m \cdot n} \right\rceil.$$

Step 1. Let P_R denote the perimeter needed to enclose a rectangular arrangement of $m \cdot n$ units of area, and let P denote the smallest amount of perimeter needed to enclose $m \cdot n$ units of area. Determine expressions for both P_R and P.

Step 2. Set $P_R = P$ and rearrange the equation to find the desired formula.

Guided Exercise 6 leads naturally to the following:

Definition 13 (Isoperimetric Pairs) Two natural numbers n and m are an **Isoperimetric Pair** if

$$m + n = \lceil 2\sqrt{m \cdot n} \rceil.$$

Guided Exercise 7 $A = m \cdot n$ **is a block number (with isoperimetric pair of divisors** n, m**) if and only if**

$$n + m = \left\lceil \frac{mn}{\lceil \sqrt{mn} \rceil} \right\rceil + \lceil \sqrt{mn} \rceil.$$

Step 1. In Guided Exercises 4 and 5, we optimally enclosed A units of area by first enclosing A^* units of area. This implies that if A is a block number provided, the rectangular block curve enclosing $A = m \cdot n$ units of area uses the same number of perimeter blocks needed to enclose a rectangular arrangement of A^* units of area.

Step 2. Use the result of Challenge Problem 8 to verify that the perimeter needed to enclose such an arrangement is given by

$$P = 2\left\lceil \frac{A}{\lceil \sqrt{A} \rceil} \right\rceil + 2\lceil \sqrt{A} \rceil + 4.$$

Step 3. Set the perimeter value in the previous step equal to the perimeter $2n + 2m + 4$, set $A = mn$, and reduce the resulting equation to obtain the desired formula.

Guided Exercise 8 Proving Theorem 3

Step 1. It suffices to verify Theorem 3 when A is neither square nor oblong. This means we may assume $A_* < A < A^*$.

Step 2. As a first case assume that A^* is a perfect square. Show that the result from Guided Exercise 7 yields the equation

$$n + m = \left\lceil \frac{mn}{\lceil \sqrt{mn} \rceil} \right\rceil + \lceil \sqrt{mn} \rceil = 2\lceil \sqrt{mn} \rceil.$$

(continued)

Step 3. Deduce from Step 2 the equation $A^* - A = \left(\lceil\sqrt{mn}\rceil - n\right)^2$, which shows that both $A^* - A$ and A^* are perfect squares.

Step 4. Now assume that both $A^* = \left\lceil\sqrt{mn}\right\rceil^2$ and $A^* - A = k^2$ are perfect squares. Conclude that $A = \left(\lceil\sqrt{mn}\rceil - k\right)\left(\lceil\sqrt{mn}\rceil + k\right) = A$.

Step 5. Verify that the divisors of A from Step 4 are, in fact, an isoperimetric pair.

Step 6. As a second and last case, assume that A^* is oblong, and conclude that

$$A^* = \left(\lceil\sqrt{mn}\rceil - 1\right)\lceil\sqrt{mn}\rceil.$$

Step 7. Conclude from Step 6 that $A = mn = 2n\lceil\sqrt{mn}\rceil - n - n^2$.

Step 8. Conclude from Step 7 that $A^* - A = \left(n - \lceil\sqrt{mn}\rceil\right)\left(n - \lceil\sqrt{mn}\rceil + 1\right)$, which is also an oblong number.

Step 9. Now assume that both A^* is an oblong number and so is $A^* - A = k(k+1)$. Conclude that

$$A = \left(\left(\lceil\sqrt{mn}\rceil - 1\right) + k\right)\left(\left(\lceil\sqrt{mn}\rceil - k\right)\right),$$

and verify that the divisors are an isoperimetric pair.

3   Additional Research Projects

3.1   New Neighbors on the Block

Research Project 1 (Corner Neighbors) Instead of defining *neighboring* or *adjacent* blocks as squares that share a common edge, one can require that two blocks *share a corner*. Using the same conventions for perimeter and area, then, what block curves solve the Block Isoperimetric Problem and the Dual Block Isoperimetric Problem? Perhaps square and "oblong" rhombic curves (such as the ones depicted in Fig. 7), but this remains to be verified.

Additional interesting questions to pursue as part of this project include the following: Are block circles unique in this setting? Is the quantity P^2/A constant on block circles? What is the value of corner-block-π? What is the sequence of corner-block numbers and how are corner-isoperimetric pairs characterized?

Fig. 7 Corner-block curves

Research Project 2 (Contact Neighbors) There is no need to be so restrictive, still, as one can define a *neighboring pair* by imposing the requirement that two squares share *at least* one point in common. This definition includes our original one, the corner-sharing one described in Research Project 1, other point-matching arrangements, and various combinations. In this scenario, the notion of enclosed area requires adjustment and the LEGO® analogy begins to fall apart (pun intended). Nonetheless, isoperimetric questions are available and remain interesting.

Research Project 3 (Alternative Blocks) Instead of or in addition to changing the definition of *neighbor*, one can change or generalize the definition of *block*. Why use only squares? Why not generalize to $1 \times n$ rectangles, general polygons, curved blocks, *L*-shaped blocks, or even general blobs? Figure 8 shows two such arrangements.

Isoperimetric and related questions about resulting notions of circles, π, block numbers, and isoperimetric pairs abound for each such choice of a new, "alternative" block.

3.2 Block Knots

It is easy to envision block-like curves that enclose multiple regions of area; some are shown in Fig. 9.

Such block arrangements appear to cross themselves and so earn the name *block knots*. More specifically, a block knot that encloses two regions of area is referred to as an *order 2 block knot* (with *order k block knots* left as hopefully obvious generalizations for the reader to contemplate as interested). As shown in Fig. 9a,b,

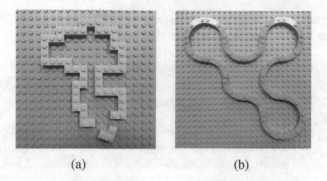

Fig. 8 Alternative block curves. (**a**) L-shaped blocks. (**b**) Circular blocks

Fig. 9 Order 2 block knots with two 3-neighbor blocks

and c, these block curves often contain two blocks with three neighbors, while Fig. 9d shows such a curve where one block has exactly four neighbors. These observations motivate the following definition.

Definition 14 (Block Knot of Order 2) A **block knot of order 2** is a collection of blocks β where every member in the collection has exactly two neighbors with two mutually exclusive exceptions: either there exist exactly two blocks that have three neighbors *or* there exists exactly one block with four neighbors.

These concepts lead naturally to Research Project 4, which is a LEGO® version of the rather famous "Double Bubble" conjecture (see [1]).

Research Project 4 (Block Double Bubble Problem) Among all order-2 block knots of a given perimeter P, which one(s) enclose(s) the maximum amount of areas $A_1 + A_2$? Among all order-2 block knots enclosing regions of fixed areas A_1 and A_2, which one(s) require(s) a minimum perimeter P?

3.3 What Is So Special About Rectangles?

Research Project 5 (Other Isoperimetric Pairs) The notions of *block numbers* and *isoperimetric pairs* introduced in Sect. 2.4 arose when determining the conditions under which *rectangular* block curves coincided with block circles, but there is no a priori need to focus on rectangles. Instead, one can take any preferred block curve shape and ask when *it* is also a block circle. How these choices impact the definition and description of Isoperimetric Pairs and Block Numbers are questions with interesting results.

3.4 Isoparametry and Blocks

This subsection does *not* contain a typo. The term "isoparametric" was coined by Apostol and Mnatsakania in [3] to describe distinct regions in the plane with equal perimeter and area. These two geometric quantities are thought of as geometric *parameters* and motivate the following:

The Isoparametric Problem Which non-congruent closed curves in the plane have the same area and perimeter?

Discretizing this question via blocks yields an obvious but equally interesting research project.

Research Project 6 (Isoparametric Block Curves) Determine those block curves that have the same perimeter, and enclose the same amount of area but are not "congruent." That is, determine when two block curves are **isoparametric**. (Here, it is natural to take "congruent" to mean that

(continued)

neither curve can be transformed into the other via translations, rotations, or reflections, but this choice can, of course, be modified.)

The classical Isoperimetric Problem and its known solutions tell us that two circles are never isoparametric, but this is not the situation in LEGO® land, as discussed in Sect. 2.4. A narrower sub-question, then, is to determine which block circles *are* isoparametric.

References

1. Alfaro, M., Brock, J., Foisy, J., Hodges, N., and Zimba, J.: The Standard Double Soap Bubble in $\mathbb{R}^2$ Uniquely Minimizes Perimeter *Pacific J. Math.*, 159, No. 1 (1993), pp. 47–59
2. Ali, The Beauties Hidden in Pi (π), Medium, (2019), Available at https://medium.com/however-mathematics/the-beauties-hidden-in-pi-%CF%80-1c614e636426 (Accessed 12/10/2020).
3. Apostol, T. and Mnatsakania, M. A.: Isoperimetric and Isoparametric Problems. *Amer. Math. Monthly*, 111 (February 2004), pp. 118–136
4. Belcastro, S.-M.,*Discrete Mathematics With Ducks*, CRC Press, 2012
5. Blasjö, V.: The Evolution of the Isoperimetric Problem. *Amer. Math. Monthly*, 112, (2001)
6. Chakerian, G. D.: The Isoperimetric Quotient: Another Look at an Old Favorite. *The College Mathematics Journal*, Vol. 22, No. 4 (Sep., 1991), pp. 313–315.
7. Hagen, C. R. and Friedman, T., Quantum Mechanical Derivation of the Wallis formula for π Journal of Mathematical Physics, 56, (2015)
8. Hammack, R., *Book of Proof*, Creative Commons, 2009.
9. Miller, S. J., Morgan, F., Newkirk, E., Pedersen L., and Seferis, D.: Isoperimetric Sets of Integers *Mathematics Magazine*, Vol. 84, No. 1 (2011), pp. 37–42
10. Santiago, S., Pi is Encoded in the Patterns of Life Biophysical Society Blog, available at https://www.biophysics.org/blog/pi-is-encoded-in-the-patterns-of-life (Accessed 1/10/21)
11. Strogatz, S., *Why Pi Matters*, New Yorker, (2015).
12. Wallen, L. J.: Kepler, the Taxicab Metric, and Beyond: An Isoperimetric Primer. *The College Mathematics Journal*, Vol. 26, No. 3 (May, 1995), pp. 178–190
13. Weigert, J.: The Sagacity of Circles: A History of the Isoperimetric Problem - Introduction. *Convergence* (2010)

Numerical Simulation of Arterial Blood Flow

Arati Nanda Pati and Kehinde O. Ladipo

Abstract

The normal human heart is a strong muscular two-stage pump which pumps continuously through the circulatory system. It is a four-chambered pump that controls blood through a series of valves in one direction. Blood flow through a blood vessel, such as vein and artery, can be modeled by Poiseuille's law which establishes a relationship between velocity of blood and the radius of the vessel. Our goal here is to solve the differential equations numerically by replacing with algebraic equations to obtain approximate solutions of velocity. We will walk through the development of the simulation procedure to visualize how blood flows in steady state. At the end of the chapter, students will be able to do projects on their own to visualize the flow simulation in one dimension.

Suggested Prerequisites *Two semesters of calculus.*

A. N. Pati (✉)
Department of Mathematics and Computer Science, University of St. Thomas, Houston, TX, USA
e-mail: patia@stthom.edu

K. O. Ladipo
Department of Mathematics, Faculty of Liberal Arts & Sciences and Innovative Learning, Humber College Institute of Technology and Advanced Learning, Toronto, ON, Canada
e-mail: kehinde.ladipo@humber.ca

1   Introduction

The human heart is a pump. It is a strong muscular organ responsible for pushing blood throughout the entire body. It consists of two side-by-side pumps—the right and the left as illustrated in Fig. 1. The passages open vertically that are separated by a muscle wall called the septum. Each side of the heart is divided into an upper chamber called an atrium and lower chamber called a ventricle. The right heart receives impure or deoxygenated blood from the veins, and the left heart receives oxygen-rich blood from the lungs. The right side pump pushes deoxygenated blood to the lungs, while the left pump pushes oxygenated blood to the rest of the body, thereby forming two separate blood circuits.

Anatomically, the heart is a four-chambered pump that controls blood flow through a series of valves in one direction. The right atrium receives the deoxygenated blood from the largest vein, called the venae cavae. Through the tricuspid valve, blood flows from the right atrium to the right ventricle. Then, through the pulmonary semilunar valve, blood flows to the lungs via pulmonary artery. In the lungs, gas exchange takes place whereby the impure blood containing carbon dioxide gets purified and filled with oxygen. Through the pulmonary vein, pure bright red blood goes to the left atrium through the mitral or bicuspid valve to the left ventricle. The left ventricular chamber is the most prominent thick muscular section; it pumps blood to the ascending aorta through the aortic semilunar valve. Although valves are passive structures, they are also responsible for preventing backflow. The

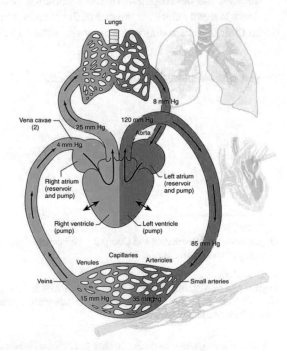

Fig. 1 Heart pump. Schematic of the systemic and pulmonary circuits. The right side of the heart pumps blood through the pulmonary circuit. The left side of the heart pumps blood via the systemic circuit to all body parts. Blood receives oxygen and loses carbon dioxide at the lungs. Freshly oxygenated blood then transported to individual cells for exchange of oxygen and waste products. Pressure difference created by the two pumps allows blood to flow in one direction only. Reproduced with Permission. https://openstax.org/books/college-physics/pages/12-4-viscosity-and-laminar-flow-poiseuilles-law [2]

left ventricle has a thick muscular wall to increase arterial blood pressure, which drives the blood flow through all parts of the body like brain to leg except lungs, called systemic circulation. The right ventricle requires less pressure, as it drains blood only to the lungs and hence has a thinner muscle wall. The left is also called the body pump and the right is called the lung pump. The largest vessel called the aorta which receives blood from the left ventricle is divided into smaller arteries and arterioles and finally to smallest capillaries where oxygen and nutrients are exchanged with the body organs such as cerebral, coronary, splanchnic, renal, and skeletal muscle. The waste products including the carbon dioxide are collected from the organs through capillaries and then transported to the right heart through venules and small and large veins. The heart pumps about 5 liters of blood per minute and approximately 7200 liters of blood per day.

The French scientist J. L. Poiseuille was interested in circulation of blood, so he put significant effort to understand the flow of blood through arteries.

Jean Léonard Marie Poiseuille was a French physician and physiologist (Fig. 2). He is famous for his unique mathematical formula which governs the fluid flow in circular tubes.

Poiseuille was born on April 22, 1799 and died on December 26, 1869 in Paris, France. His father was Jean Baptiste Poiseuille, a carpenter, and his mother was Anne Victoire Caumont. He married to De la Lorette in 1829. When he was 18, he went to the École Polytechnique to study physics and mathematics. Unfortunately, the school was closed after a year for political reasons, and when it reopened, Poiseuille did not return. Instead, he chose to study medicine. Poiseuille received his medical degree in 1828. During his doctoral research on "The Force of the Aortic Heart," Poiseuille invented the U-tube mercury manometer, the hemodynamometer, which he used to measure blood pressure in the arteries of horses and dogs. Later, the instrument was named Poiseuille–Ludwig hemodynamometer. It was recording the

Fig. 2 Jean Léonard Marie Poiseuille. Source: https://en. wikipedia.org

results in graphic form in real time. This device was in practice in medical schools until the 1960s. To this day, blood pressures are measured in millimeter of mercury (mm Hg) as the standard unit of measurement of pressure due to Poiseuille's invention.

Poiseuille's interest in blood circulation led him to conduct a series of experiments independently on flow of liquids in narrow tubes. All of Poiseuille's tests were carried out on horizontal tubes. His experimental insights were expressed in terms of a mathematical formula which is permanently used in the medical field. This equation states that the flow rate is determined by the viscosity of the fluid, the pressure drop along the tube, and the tube diameter. Although he derived the formula in 1838, it was not published until 1846. Eduard Hagenbach coined this formula as Poiseuille's law in 1860. Gotthilf Heinrich Ludwig Hagen, a German hydraulic engineer discovered the same formula independently. Therefore, it is also known as the Hagen–Poiseuille law. Poiseuille's research on physiology won him the Montyon Medal four times and membership in the Académie de Médicine. In Poiseuille's honor, Deeley and Parr proposed naming the C.G.S. unit of viscosity the "Poise" in 1913. Poiseuille's law is one of the few equations that is still well known in the modern twenty-first century cutting edge medical field. Since his first discovery, Poiseuille's influence has extended vastly and is significantly visible in almost all fields of sciences.

Poiseuille's law describes the relation between pressure drop and the blood flow rate under steady flow condition. Poiseuille's law can be derived from basic principles of physics or general Navier–Stokes equations. The first derivation was done by Eduard Hagenbach from the Navier–Stokes equations in 1860. Poiseuille flow is called the laminar or swirl-free flow. In laminar flow, fluid flow is smooth. It flows in layers describing parallel paths as shown in Fig. 3. The tube is considered to be solid, straight, and uniform. The pipe is long enough to allow the Poiseuille flow to be fully developed or steady. That is, it does not change with time and stays

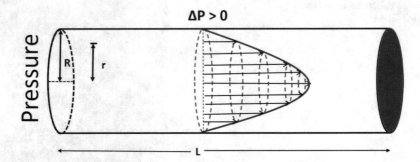

Fig. 3 Blood flow in an artery. Due to pressure difference, blood flows in layers. Therefore, it is called laminar flow. The velocity profile is parabolic where it is maximum at the center and zero at the wall. The radius of the artery is R, length is L, and pressure difference between inlet and outlet is ΔP. Here, r is a variable given by $0 < r < R$ representing center of the artery to the wall

unchanged or steady. The velocity at the wall is zero. These properties are based on the following three major assumptions of Poiseuille's law:

- The tube is stiff, straight, and uniform.
- Blood is Newtonian, i.e., viscosity is constant.
- The flow is laminar and steady and velocity at the wall is zero.

Viscosity is the property of the fluid which we have discussed below. Generally, fluid flows from a region of higher pressure to a region of lower pressure. With laminar and steady flow, the velocity profile is a parabola. Hagenbach derived this parabolic expression for velocity distribution for the first time. When fluid flows through a uniform tube with radius R, the velocity profile over the cross section is a parabola as shown in Fig. 3. It is given by the following formula as a function of radius, r:

$$V(r) = \frac{\Delta P}{4 \eta L} \left(R^2 - r^2 \right),$$ (1)

where ΔP is the pressure drop in the tube and r varies between 0 and R. This is the difference between the pressure at the entrance and exit of the tube of length L and η is the viscosity of fluid. The velocity $V(r)$ is a quadratic function of r, so the graph is a parabola. We observe that at the axis, when $r = 0$, velocity is maximum and at the wall, when $r = R$, velocity is zero. Evaluating (1) at $r = 0$, the maximum velocity can be expressed by

$$V_{max} = \frac{\Delta P}{4 \eta L} R^2.$$ (2)

Hence, velocity shown in (1) can be rewritten as

$$V(r) = V_{max} \left(1 - \frac{r^2}{R^2} \right).$$ (3)

Since velocity $V(r)$ is a variable velocity that changes between 0 and V_{max}, we can find the average or mean velocity. In medical applications, volume flow rate plays an important role. The formula, which is very well known in the medical community, can be derived from a complex set of mathematical equations related to mean velocity in a tube. Without proof, we will quote the formula as follows:
The mean or average velocity is given by

$$V_{mean} = \frac{\Delta P \, R^2}{8 \eta L}.$$ (4)

If we compare (2) and (4), we observe that $V_{mean} = \frac{1}{2}V_{max}$. The mean velocity is attained at $r = \frac{1}{\sqrt{2}}R \approx 0.707\ R$. We will represent this location on the radius by r^* as listed in Table 2. In Fig. 12a, we can visualize V_{max} and V_{mean} at their corresponding locations.

Blood flow rate Q is obtained by multiplying the mean velocity with the cross-sectional area of the tube, πR^2,

$$Q = \frac{\pi\ \Delta P\ R^4}{8\ \eta\ L}. \tag{5}$$

This Eq. (5) is the famous **Poiseuille's law**. It describes the relation between pressure drop, ΔP, and the steady fluid flow rate Q through the constant radius and solid straight tube. The flow rate Q also depends on η, the viscosity, L, the length of the cylindrical tube, and R, the radius of the tube. Q is directly proportional to ΔP and fourth power of R and inversely proportional to η and L. The terms "velocity" in (1) and "flow" in (5) should not be confused. Velocity is the distance an object moves with respect to time (i.e., the distance traveled per unit of time). In contrast, flow is the volume of fluid that is moving per unit of time. The mathematical expression for velocity in a circular tube is a quadratic function best known as a parabola. It represents varying velocities at different locations of the radius such that it is maximum at the center and decreasing to zero at the boundary. Therefore, in practical application of blood flow, the average or mean velocity given by (4) and measured by Doppler flow meter is used. The significance of Poiseuille's law (5) is discussed below after Example 1.

The viscosity, η, is the property of the fluid which refers to thickness or stickiness of the fluid. Viscosity is a measure of the resistance to flow. It varies from one fluid to other. For example, honey is stickier than water and hence harder to flow. Viscosity also depends on temperature. Therefore, with increase of temperature, honey can be made to flow faster. The units of viscosity are $Pa \cdot s = (N/m^2) \cdot s$ or Poise   (dynes $\cdot$ s/cm^2) with 1 $Pa \cdot s = 10$ Poise. Viscosity η is sometimes called dynamic viscosity. On the other hand, when viscosity is divided by density ρ, thus, $v = \eta/\rho$ is called kinematic viscosity. Blood viscosity is also a measure of resistance of blood to flow. At normal level, blood viscosity is given by

$$\eta = (3 \sim 4) \cdot 10^{-3}\ Pa \cdot s$$

$$v = \frac{\eta}{\rho} = \frac{(3 \sim 4) \cdot 10^{-3}\ Pa \cdot s}{1.06 \cdot 10^3\ Kg/m^3} = (2.8 \sim 3.8) \cdot 10^{-6}\ \frac{m^2}{s},$$

where 1 Pa $= 1\ N/m^2 = 1\ Kg/m \cdot s^2$.

There are many applications of formulas (1)–(5) available in the literature. We consider a few examples that will illustrate the impact of these formulas.

Example 1 Consider a fluid flows through a cylindrical tube. Find the flow rate of the fluid under the following conditions:

(a) If the pressure difference between the ends of the tube is doubled.
(b) If the radius of the tube is doubled.
(c) If the radius of the tube is halved.
(d) If the viscosity is halved.
(e) If the length of the cylindrical tube is doubled.
(f) If the length of the cylindrical tube is halved.

Solution We will introduce Q_N to be the new flow rate in all the parts below:

(a) Let Q_N be the new flow rate due to change in pressure difference from ΔP to $2\,\Delta P$. From (5), we obtain

$$
\begin{aligned}
Q_N &= \frac{\pi\,(2\,\Delta P)\,R^4}{8\,\eta\,L} \\[2mm]
&= \frac{2\,\pi\,(\Delta P)\,R^4}{8\,\eta\,L} \\[2mm]
&= 2\frac{\pi\,(\Delta P)\,R^4}{8\,\eta\,L} \\[2mm]
&= 2\,Q.
\end{aligned}
\tag{6}
$$

So, the new flow rate will be doubled.

(b) Let Q_N be the new flow rate due to change in radius from R to $2R$. From (5), we obtain

$$
\begin{aligned}
Q_N &= \frac{\pi\,(\Delta P)\,(2\,R)^4}{8\,\eta\,L} \\[2mm]
&= \frac{\pi\,(\Delta P)\,16\,R^4}{8\,\eta\,L} \\[2mm]
&= 16\frac{\pi\,(\Delta P)\,R^4}{8\,\eta\,L} \\[2mm]
&= 16\,Q.
\end{aligned}
\tag{7}
$$

So, the new flow rate will be 16 times more than the original flow rate.

(c) Let Q_N be the new flow rate due to change in radius from R to $\frac{R}{2}$. From (5), we obtain

$$Q_N = \frac{\pi \ (\Delta P) \ \left(\frac{R}{2}\right)^4}{8 \, \eta \, L}$$

$$= \frac{\pi \ (\Delta P) \ \frac{R^4}{16}}{8 \, \eta \, L} \tag{8}$$

$$= \frac{1}{16} \frac{\pi \ (\Delta P) \ R^4}{8 \, \eta \, L}$$

$$= \frac{1}{16} \, Q.$$

So, the new flow rate will be $\frac{1}{16}$ times less than the original flow rate.

(d) Let Q_N be the new flow rate due to change in viscosity from η to $\frac{\eta}{2}$. From (5), we obtain

$$Q_N = \frac{\pi \ (\Delta P) \ R^4}{8 \, \frac{\eta}{2} \, L}$$

$$= 2 \frac{\pi \ (\Delta P) \ R^4}{8 \, \eta \, L} \tag{9}$$

$$= 2 \, Q.$$

So, the new flow rate will be 2 times more than the original flow rate when viscosity is reduced to half.

(e) Let Q_N be the new flow rate due to change in length from L to $2L$. From (5), we obtain

$$Q_N = \frac{\pi \ (\Delta P) \ R^4}{8 \, \eta \, (2L)}$$

$$= \frac{1}{2} \frac{\pi \ (\Delta P) \ R^4}{8 \, \eta \, L} \tag{10}$$

$$= \frac{1}{2} \, Q.$$

So, the new flow rate will be reduced to half of the original flow rate when length of the pipe is doubled.

(f) Let Q_N be the new flow rate due to change in length from L to $\frac{L}{2}$. From (5), we obtain

$$Q_N = \frac{\pi \, (\Delta P) \, R^4}{8 \, \eta \, \left(\frac{L}{2}\right)}$$

$$= 2 \frac{\pi \, (\Delta P) \, R^4}{8 \, \eta \, L} \tag{11}$$

$$= 2 \, Q.$$

So, the new flow rate will be doubled the original flow rate when length of the pipe is halved.

Significance of Poiseuille's Law Let us understand the importance of Poiseuille's equation given by (5) and the impact of the flow rate in medical applications. We observe that flow rate Q is directly proportional to the fourth power of radius R. From Example 1, we see that doubling the radius will increase the flow rate by a factor of 16. Similarly, if the radius is halved, then the flow rate will be reduced by a factor of $\frac{1}{16}$. In human body, carotid arteries supply blood to the upper part, for example, to the brain as shown in Fig. 4. Due to plaque buildup on the artery walls caused by the thickening or hardening of the blood vessels of the artery due to fats, cholesterol, and other substances, the cross-sectional radius of the artery reduces, which leads to decrease in blood flow rate. So, blood carrying oxygen will be reduced. To maintain the flow rate, the heart has to work harder. This is a medical

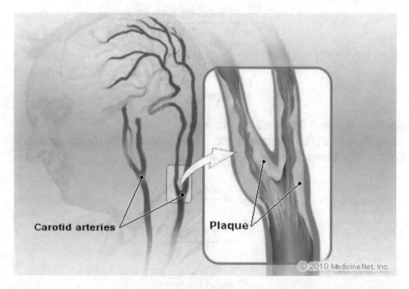

Fig. 4 Atherosclerosis in artery which reduces the radius of the artery. Atherosclerosis or stenosis is a disease of the heart called cardiovascular disease that refers to the thickening or hardening of the blood vessels of the artery due to fats, cholesterol, and other substances on the artery walls. Reproduced with permission. Source: https://Medicinenet.com [21]

condition called atherosclerosis or stenosis which is a heart disease that can lead to death. To ensure that a normal flow rate is attained while keeping pressure difference and radius unchanged, alternatively, we can decrease the coefficient of viscosity η. Sometimes doctors prescribe blood thinner medications to dilute blood and reduce blood viscosity as well as lower the blood pressure while preventing blood clots formation.

Example 2 Consider blood flows through an artery. Determine the percentage change of blood flow rate under the following conditions:

(a) Suppose with exercise, the artery increases in radius by 10%.
(b) Suppose plaque builds up due to cholesterol which causes 20% decrease in the radius of the artery.
(c) Suppose with medication, the viscosity of blood is reduced by 15%.

Solution Let Q_N be the new flow rate.

(a) Since there is a 10% increase in the radius of the artery, the new radius will be $1.10\ R$. The blood flow rate will be given by

$$Q_N = \frac{\pi\ (\Delta P)\ (1.10\ R)^4}{8\ \eta\ L}$$

$$= \frac{\pi\ (\Delta P)\ 1.46\ R^4}{8\ \eta\ L} \tag{12}$$

$$= 1.46 \frac{\pi\ (\Delta P)\ R^4}{8\ \eta\ L}$$

$$= 1.46\ Q.$$

So, there is a 46% increase in the flow rate.

(b) Since there is a 20% decrease in the radius of the artery, the new radius will be $0.80\ R$. The blood flow rate will be given by

$$Q_N = \frac{\pi\ (\Delta P)\ (0.8R)^4}{8\ \eta\ L}$$

$$= \frac{\pi\ (\Delta P)\ 0.41R^4}{8\ \eta\ L} \tag{13}$$

$$= 0.41 \frac{\pi\ (\Delta P)\ R^4}{8\ \eta\ L}$$

$$= 0.41Q.$$

So, there is a 59% decrease in the flow rate.

(c) Since there is a 15% decrease in the viscosity, the new viscosity will be $0.85\,\eta$

$$Q_N = \frac{\pi\,(\Delta P)\,R^4}{8\,(0.85\eta)\,L}$$

$$= \frac{1}{0.85}\,\frac{\pi\,(\Delta P)\,R^4}{8\,\eta\,L} \tag{14}$$

$$= \frac{1}{0.85}\,Q$$

$$= 1.18\,Q.$$

So, there is a 18% increase in the flow rate.

Example 3 Consider blood flows through an artery. Use Poiseuille's law to calculate blood flow rate for the following values: $\eta = 0.027$ dynes $\cdot$ s/cm^2, $R = 0.006$ cm, $L = 2$ cm, and $\Delta P = 5000$ dynes/cm^2.

Solution

$$Q = \frac{\pi\,\Delta P\,R^4}{8\,\eta\,L}$$

$$= \frac{\pi\,(5000\text{ dynes/cm}^2)\,(0.006\text{ cm})^4}{8\,(0.027\text{ dynes}\cdot\text{s/cm}^2)\,(2\text{ cm})} \tag{15}$$

$$= 4.71 \times 10^{-5}\text{ cm}^3/\text{s}.$$

Therefore, 4.71×10^{-5} cm^3/s volume of blood passes a point in the artery in every second.

Example 4

(a) Use Poiseuille's law to find pressure difference. Suppose water has viscosity $\eta = 0.001$ Pa $\cdot$ s at 20 °C. What pressure difference is necessary to keep water moving with a volume flow rate $Q = 0.025$ m^3/s through a 2 m long pipe with a radius $R = 3$ cm?

(b) Suppose water is flowing from left to right in a pipe as shown in Fig. 5 with length 10 m. Suppose pressure at the entrance A is 5000 Pa when $L = 0$ m. Use part (a) to find pressures at the points B, C, D, E, and F which are located 2 m apart as we move from left to right.

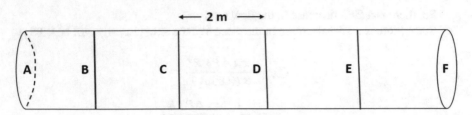

Fig. 5 Pressure on pipe

(c) What is the pressure difference between points A and F? Repeat part (a) to find pressure difference for length $L = 10$ m. Compare the answers and explain what you observe.

Solution

(a) Using Poiseuille's law (5), we rewrite the equation for pressure difference.

$$Q = \frac{\pi \; \Delta P \; R^4}{8 \; \eta \; L}$$

$$\Delta P = \frac{8 \; Q \; \eta \; L}{\pi \; R^4}.$$

(16)

Since 3 cm = 0.03 m, we get the following pressure difference:

$$\Delta P = \frac{8 \; Q \; \eta \; L}{\pi \; R^4}$$

$$= \frac{8 \; (0.025 \; \text{m}^3/\text{s}) \; (0.001 \; \text{Pa} \cdot \text{s}) \; 2 \; \text{m}}{\pi \; (0.03 \; \text{m})^4}$$

(17)

$$= 157.19 \; \text{Pa}$$

$$\approx 157 \; \text{Pa}.$$

Solution

(b) We have pressure $P = 5000$ Pa at A that corresponds to length $L = 0$ m. Since $\Delta P = 157$ Pa, there will be a pressure drop at every 2 m along the pipe as we move from left to right. Here, B corresponds to length $L = 2$ m, C corresponds to length $L = 4$ m, D corresponds to length $L = 6$ m, E corresponds to length $L = 8$ m, and F corresponds to length $L = 10$ m or the end of the pipe in Fig. 5. Therefore, we have the following pressures at the corresponding points:

$$A = 5000 \text{ Pa}$$
$$B = 4843 \text{ Pa}$$
$$C = 4686 \text{ Pa}$$
$$D = 4529 \text{ Pa} \tag{18}$$
$$E = 4372 \text{ Pa}$$
$$F = 4215 \text{ Pa.}$$

We obtained the pressure difference between A and F or at entrance and exit is $5000 - 4215 = 785$ Pa. We could also get the same result if we consider part (a) with length $L = 10$ m.

$$\Delta P = \frac{8\,Q\,\eta\,L}{\pi\,R^4}$$

$$= \frac{8\,(0.025 \text{ m}^3/\text{s})\,(0.001 \text{ Pa}\cdot\text{s})\,10 \text{ m}}{\pi\,(0.03 \text{ m})^4} \tag{19}$$

$$= 785.9503363 \text{ Pa}$$

$$\approx 786 \text{ Pa} \qquad \text{with some rounding error.}$$

Example 5 Intravenous (IV) fluid. A patient receives intravenous (IV) fluid at the rate of 0.250 mL/s through a needle with radius 0.300 mm and a length of 2.00 cm. The needle is filled with a solution of viscosity 2.00×10^{-3} Pa·s. The gauge pressure of the blood in the patient's vein is 19 mm Hg. What pressure is needed at the entrance of the needle in order to inject the solution?

$$\Delta P = \frac{8\,L\,\eta\,Q}{\pi\,R^4}$$

$$= \frac{8\,(0.02 \text{ m})\,(2 \times 10^{-3} \text{ Pa}\cdot\text{s})}{\pi\,(0.0003 \text{ m})^4} \cdot \left(2.5 \times 10^{-7}\frac{\text{m}^3}{\text{s}}\right) \tag{20}$$

$$= 3144 \text{ Pa.}$$

Since $\Delta P = P_1 - P_2$ is the change in pressure, where P_1 is the pressure at the entrance and P_2 is the gauge pressure of blood at patient's vein, we have

$$P_1 = \Delta P + P_2.$$

Converting 19 mm Hg = 2533 Pa at the vein, we obtained

$$P_1 = 3144 + 2533 = 5677 \text{ Pa.}$$

Therefore, we need 5677 Pa, pressure at the entrance of the needle in order to inject the intravenous solution.

Example 6 Consider a blood vessel with radius 0.1 cm, length 6 cm, pressure difference 1000 dynes/cm^2, and viscosity $\eta = 0.027$ dynes $\cdot$ s/cm^2.

(a) Find the velocity of the blood along the center line $r = 0$, at radius $r = 0.02$ cm, $r = 0.04$ cm, $r = 0.06$ cm, and $r = 0.08$ cm, and at the wall $r = R = 0.1$ cm.
(b) Find the velocity of the blood at radius $r = -0.02$ cm, $r = -0.04$ cm, $r = -0.06$ cm, and $r = -0.08$ cm and at the wall $r = R = -0.1$ cm, and discuss your results with (a).
(c) Find the mean velocity and maximum velocity.

Solution We have

$$V(r) = \frac{\Delta P}{4 \eta L} \left(R^2 - r^2 \right)$$

with $R = 0.1$ cm, $L = 6$ cm, $P = 1000$ dynes/cm^2, and $\eta = 0.027$ dynes $\cdot$ s/cm^2. Hence, the velocity $V(r)$ will be expressed as

$$V(r) = \frac{1000 \text{ dynes/cm}^2}{4 \, (0.027 \text{ dynes} \cdot \text{s/cm}^2) \cdot 6 \text{ cm}} \left(0.1^2 \text{ cm}^2 - r^2 \right). \tag{21}$$

Therefore, we obtain the following velocities at given locations along the radius:

$V(0) = 15.432$ cm/s
$V(0.02) = 14.814$ cm/s
$V(0.04) = 12.9629$ cm/s
$V(0.06) = 9.876$ cm/s
$V(0.08) = 5.55$ cm/s
$V(0.1) = 0$ cm/s

(b) We will obtain the following velocities at different values of r:

$V(-0.02) = 14.814$ cm/s
$V(-0.04) = 12.9629$ cm/s
$V(-0.06) = 9.876$ cm/s
$V(-0.08) = 5.55$ cm/s
$V(-0.1) = 0$ cm/s

In Table 1, we have listed the velocities at different values of radius r. We have shown the velocity lines in Fig. 6a and the parabolic velocity profile in Fig. 6b. In

Table 1 Velocity at discrete points

Radius (cm)	Velocity (cm/s)
0.1	0
0.08	5.55
0.06	9.876
0.04	12.963
0.02	14.814
0	15.432
−0.02	14.814
−0.04	12.963
−0.06	9.876
−0.08	5.55
−0.1	0

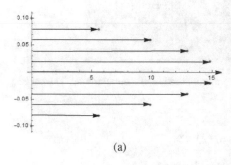

(a)

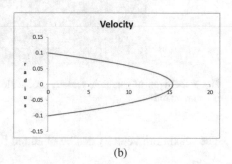

(b)

Fig. 6 Velocity profile. (**a**) Velocity lines. (**b**) Parabolic velocity

3D circular tube, the velocity profile is a paraboloid, a parabolic surface which is shown in Fig. 7.

(c) The mean velocity will be given by (4). Hence,

$$V_{mean} = \frac{(1000 \text{ dynes/cm}^2) \cdot 0.1^2 \text{ cm}^2}{8 \,(0.027 \text{ dynes} \cdot \text{s/cm}^2) \cdot 6 \text{ cm}} \tag{22}$$
$$= 7.716 \text{ cm/s}.$$

Alternatively, the mean velocity can also be calculated from (1) at $r = \frac{R}{\sqrt{2}} = \frac{0.1}{\sqrt{2}}$ cm.

$$V\left(\frac{0.1}{\sqrt{2}}\right) = \frac{1000 \text{ dynes/cm}^2}{4 \,(0.027 \text{ dynes} \cdot \text{s/cm}^2) \cdot 6 \text{ cm}} \left(0.1^2 \text{ cm}^2 - \left(\frac{0.1}{\sqrt{2}}\right)^2 \text{ cm}^2\right)$$

$$= 7.716 \text{ cm/s}. \tag{23}$$

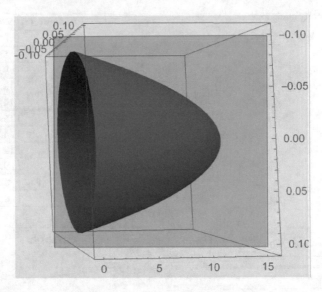

Fig. 7 3D view of parabolic velocity profile along a cross section of a circular tube. The surface is a paraboloid

The maximum velocity can be obtained from (2). So

$$V_{max} = \frac{(1000 \text{ dynes/cm}^2) \cdot 0.1^2 \text{ cm}^2}{4 \, (0.027 \text{ dynes} \cdot \text{s/cm}^2) \cdot 6 \text{ cm}} \tag{24}$$

$$= 15.432 \text{ cm/s}.$$

We observe that $V_{mean} = \frac{1}{2} V_{max}$.

We have seen some application examples of Poiseuille's equations (1)–(5) in a 3D circular tube. However, the real-world problems are very complex. Sometimes, researchers and scientists use advanced mathematical modeling and computational simulation techniques. In order to understand the basic flow pattern, our goal here is to reduce the flow problem to a very simplified one dimensional problem. Therefore, our plan for the next sections is as follows. In Sect. 2, we will derive all the previous equations for a simpler planar model for Poiseuille flow in a 2D channel. In Sect. 3, we will apply integration to solve the 2D model and compare the results with the 3D equations presented in Sect. 1. Ultimately, the 2D model will be reduced to 1D problem which will be solved using numerical techniques. The resulting system of linear equations will be solved by hand, EXCEL, and MATLAB in Sect. 4. Finally, we will discuss some exercises in Sects. 5 and 6 and two research projects in Sect. 7 to gain insight into the fundamental understanding of blood flow.

2   Planar Model of Poiseuille Flow in 2D Channel

Poiseuille flow models the steady flow in a tube with circular cross section as shown in Fig. 8. Consider a horizontal slice along the diameter of the 3D tube to obtain a 2D channel as shown in Fig. 9. This is analogous to flow between two stationary parallel plates. The flow is induced by difference in pressure between the two sections of the channel. Fluids flow from region of high pressure to region of low pressure. In order to develop the simplified 2D model of Poiseuille flow, we will assume that:

- The axis of the channel is along the x-axis at $y = 0$, and the flow is in one direction parallel to the x-axis.
- The flow through the section from $x = 0$ to $x = L$ is steady, and it is between the two stationary parallel plates (walls) at a distance $2H$ apart, located at $y = -H$ and $y = H$ (See Fig. 9).
- At each cross section of the channel, the velocity V is maximum at the center and 0 at the walls. The velocity can therefore be described approximately by a function of only one variable, y as $V = V(y)$.
- The viscosity η is constant.
- We denote the pressure at the left by P_1 and at the right by P_2, with $P_1 > P_2$, so that $\Delta P = (P_1 - P_2) > 0$ is a constant and the flow is from left to right.

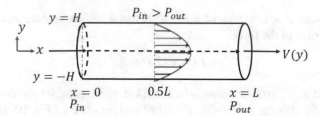

Fig. 8 Fluid through a circular pipe. The axis of the pipe is at $y = 0$. The length of the pipe is L and width is $2H$. Due to pressure difference at inlet and outlet, fluid flows from left to right. The velocity is maximum at the center and zero at the walls

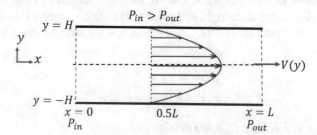

Fig. 9 Fluid flow between two fixed plates. The axis of the channel is at $y = 0$. The length of the channel is L and width is $2H$. Due to pressure difference at inlet and outlet, fluid flows from left to right. The velocity is maximum at the center and zero at the walls

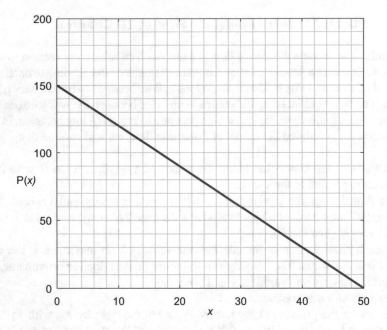

Fig. 10 Graph of pressure $P(x) = -30x + 150$ as a linear function

The pressure P at any point of the channel can therefore be described by a linear function of x of the form:

$$P(x) = Ax + B, \tag{25}$$

where A and B are arbitrary constants. For illustration, consider the linear function $P(x)$ given in (25) for $A = -30$, $B = 150$ and depicted in Fig. 10. Observe that the slope is $A = -30$, and the y-intercept is $(0, 150)$. This demonstrates that the derivative of $P(x)$ is a constant, establishing that it is a linear function.

Now let us solve for the coefficients A and B in Eq. (25). We notice that at the left end point of the channel when $x = 0$, the inlet pressure can be expressed as $P(0) = P_1$. Similarly, at the right end point of the channel when $x = L$, the outlet pressure can be expressed as $P(L) = P_2$. When we substitute these into (25), we get

$$P(0) = P_1 = B$$
$$P(L) = P_2 = AL + B. \tag{26}$$

Solving for A and B from (26), we obtain

$$A = \frac{(P_2 - P_1)}{L} \tag{27}$$

$$B = P_1. \tag{28}$$

Hence, (25) can be expressed in terms of the inlet and outlet pressures as

$$P(x) = \frac{(P_2 - P_1)}{L}x + P_1.$$ (29)

In terms of the pressure difference $\Delta P = (P_1 - P_2)$, the equation becomes

$$P(x) = -\frac{\Delta P}{L}x + P_1.$$ (30)

The slope is the derivative of $P(x)$ which is

$$\frac{dP}{dx} = -\frac{\Delta P}{L}.$$ (31)

The Poiseuille model of the arterial blood flow is described by a second-order ordinary differential equation (ODE). The ODE originates from Newton's second law of motion which requires that the rate of change of momentum equals the applied force. In this simplified model, we assume that the only two forces acting on the fluid particles are the net pressure force $\left(-\frac{dP}{dx}\right)$ and the net viscous force $\eta \frac{d^2V}{dy^2}$ per unit area of fluid particle. The derivation of the ODE is beyond the scope of our discussion, and we will therefore quote it without proof as

$$0 = \left(\eta \frac{d^2V}{dy^2}\right) + \left(-\frac{dP}{dx}\right)$$ (32)

or

$$\eta \frac{d^2V}{dy^2} = \frac{dP}{dx}.$$ (33)

The known velocities at the walls, called the boundary conditions of the ODE, are given by

$$V = 0 \text{ on the walls of the channel at } y = -H \quad \& \quad y = H.$$ (34)

If we substitute the derivative of $P(x)$ from (31), we get

$$\frac{d^2V}{dy^2} = -\frac{\Delta P}{\eta L}.$$ (35)

3   Exact Solution of the 2D Planar Model

The exact solution $V(y)$ of (35), also called the analytic solution, can be derived by integrating the equation with respect to y twice. Integrating (35), first, we get

$$\int \frac{d}{dy}\left(\frac{dV}{dy}\right) dy = \int -\frac{\Delta P}{\eta L} \, dy \tag{36}$$

$$\frac{dV}{dy} = -\frac{\Delta P}{\eta L} y + C, \tag{37}$$

where C is an arbitrary constant. When we integrate (37) with respect to y again, we get

$$\int \frac{dV}{dy} \, dy = \int \left(-\frac{\Delta P}{\eta L} y + C\right) dy \tag{38}$$

$$V(y) = -\frac{\Delta P}{2\eta L} y^2 + Cy + D, \tag{39}$$

where D is a second arbitrary constant. The next step is to find the values of C and D. Applying the boundary conditions at both walls in (34), it leads to

$$0 = V(H) = -\frac{\Delta P}{2\eta L} H^2 + CH + D$$

$$0 = V(-H) = -\frac{\Delta P}{2\eta L} H^2 - CH + D. \tag{40}$$

Adding the above equations, we get

$$-2 \frac{\Delta P}{2\eta L} H^2 + 2D = 0. \tag{41}$$

Therefore,

$$D = \frac{\Delta P \, H^2}{2 \, \eta \, L}. \tag{42}$$

Substituting the value of D into the first part of (40), we obtain

$$0 = -\frac{\Delta P}{2\eta L} H^2 + CH + \frac{\Delta P}{2 \, \eta \, L} H^2$$

$$0 = C H. \tag{43}$$

Since $H \neq 0$, it leads to

$$C = 0. \tag{44}$$

Therefore, the analytic or exact solution of the ODE (35) in terms of the pressure drop is

$$V(y) = \frac{\Delta P}{2\eta L}\left(H^2 - y^2\right), \tag{45}$$

which is a parabola. We can factor out H^2 to get the equation for the velocity for planar Poiseuille flow as

$$V(y) = \frac{\Delta P H^2}{2\eta L}\left(1 - \frac{y^2}{H^2}\right). \tag{46}$$

A parabola is a geometric figure which consists of points that are equidistant from a fixed point called the *focus* and a fixed line called the *directrix*. Since the vertex is a particular point on the parabola, its distance from the focus must also be equal to its distance from the directrix. A parabola has an *axis of symmetry* which may be along the x-axis or along the y-axis. The axis of symmetry passes through the vertex and the focus and intersects the directrix. The common distance from the vertex to the focus and directrix is denoted by p, and the vertex is denoted by the point (h, k). With these notations, the *standard equation* of a parabola is

$$(y - k)^2 = 4p\,(x - h) \text{ if axis of symmetry is along the } x\text{-axis} \tag{47}$$

$$(x - h)^2 = 4p\,(y - k) \text{ if axis of symmetry is along the } y\text{-axis.} \tag{48}$$

For a detailed discussion on parabolas, the reader is referred to High School Geometry, College Algebra, or Pre-calculus textbooks. Also, there are now a lot of materials on the Internet that brilliantly provide a detailed discussion on parabolas.

In the example that follows, we will demonstrate that any function of the form (45) represents a parabola, and we will determine the vertex, focus, and directrix.

Example 7 Consider the function $V(y)$ in (45), where $y = -H \leq y \leq H$. For simplicity, assume that the coefficient $C = \frac{\Delta P}{2\eta L}$ is a constant, and set $C = 1$ and $H = 2$.

(a) Plot the ordered pairs $(V(y), y)$, where $x = V(y)$. Observe that the graph represents a parabola with vertex coinciding with the point where the maximum of $V(y)$ occurs.
(b) Write the coordinates (h, k) of the vertex.

(c) Write the standard equation of the parabola.
(d) Determine the coordinates of the focus and the equation of the directrix.

Solution When $C = 1$ and $H = 2$, we get the function of y, $V(y) = (4-y^2)$ whose graph is shown in Fig. 11. From the figure, the maximum value of $V(y)$ occurs at point $(4, 0)$ when $y = 0$. So, this is a parabola with vertex at $(h, k) = (4, 0)$. This clearly demonstrates that any function of the form (45) represents a parabolic function.

Using notations p and (h, k) as described above, the standard equation of the parabola with axis of symmetry along the x axis is

$$(y - 0)^2 = C(x - 4),$$

where $k = y = 0$ and $h = V(0) = 4$.

The common distance p from the focus and directrix to the vertex can be determined by setting $4p = C$. In this example where $C = 1$, we get $p = 0.25$.

To determine the coordinates of the focus and the equation of the directrix, we draw a horizontal line in Fig. 11 through the vertex. From the vertex, measure along the horizontal line, a length of $p = 0.25$ units to the right and the left of the vertex. The point p units to the right is on the directrix, and the point p units to the left is

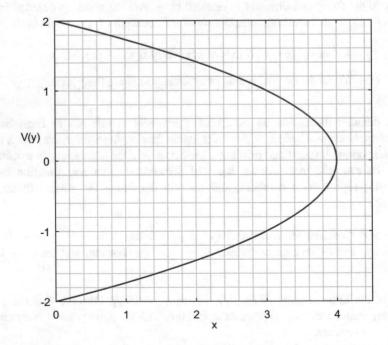

Fig. 11 Graph of $V(y) = C(H^2 - y^2)$ when $C = 1$, $y = -2 \le y \le 2$ as in Example 7

the focus. The focus at distance p to the left of the vertex is the point $(h - p, k)$ or point $(3.75, 0)$ in this case. The directrix at distance p to the right of the vertex is the vertical line $x = h + p$ or line $x = 4.25$ in this case.

From algebra, we know that quadratic functions give rise to parabolas. This simple mathematical tool is used for many real-life applications. In the seventeenth century, Galileo discovered the trajectory of a particle to be parabolic. There are many suspension bridges with parabolic shapes including the steel arch of Rainbow Bridge on Niagara River that connects the USA and Canada since 1941. It is also used in highway engineering to construct vertical curve roads. Dish antenna and parabolic microphones have parabolic reflectors. There are many more applications. In our current discussion, we see how a simple mathematical function (45) is used in blood flow problems.

In the following discussion, we will calculate the maximum velocity (V_{max}) and average or mean velocity (V_{mean}) and the locations in the channel where they occur. We will also calculate the flow rate (Q) for the planar 2D flow channel. From Calculus, we know that first derivative plays an important role in finding minimum and maximum of a function at the *critical points*. We denote the first derivative of $V(y)$ by $V'(y)$ or $\frac{dV}{dy}$. From Example 7, we have seen that the vertex is a *critical point* where the maximum velocity V_{max} occurs and the first derivative $V'(y) = 0$. This means

$$V'(y) = \frac{\Delta P}{2\eta L} (-2y) = 0. \tag{49}$$

Therefore, V_{max} occurs at the vertex $y = 0$. We will derive the equation for V_{max} by substituting $y = 0$ into Eq. (45) to get

$$V_{max} = V(0) = \frac{\Delta P H^2}{2\eta L}. \tag{50}$$

Hence, the analytic solution (46) can be expressed in terms of the maximum velocity V_{max} as

$$V(y) = V_{max} \left(1 - \frac{y^2}{H^2}\right). \tag{51}$$

The volumetric flow rate Q of a fluid across a confined channel is defined by

$$Q = 2H \times V_{mean}. \tag{52}$$

An alternative formula for Q is obtained by integrating $V(y)$ with respect to y across the channel as follows:

$$Q = \int_{-H}^{H} V(y)dy = \int_{-H}^{H} \frac{\Delta P H^2}{2\eta L} \left(1 - \frac{y^2}{H^2}\right) dy \tag{53}$$

$$= \frac{\Delta P H^2}{2\eta L} \left[y - \frac{y^3}{3H^2}\right]_{-H}^{H} \tag{54}$$

$$= \frac{\Delta P H^2}{2\eta L} \left[\left(H - \frac{H^3}{3H^2}\right) - \left(-H - \frac{(-H)^3}{3H^2}\right)\right] \tag{55}$$

$$= \frac{\Delta P H^2}{2\eta L} \left[\left(H - \frac{H}{3}\right) - \left(-H + \frac{H}{3}\right)\right] \tag{56}$$

$$= \frac{\Delta P H^2}{2\eta L} \left[\frac{4H}{3}\right] \tag{57}$$

$$Q = \frac{2\Delta P H^3}{3\eta L}. \tag{58}$$

In a similar manner to 3D tube, we will derive the average or mean velocity for a 2D channel. In this case, V_{mean} is given by

$$V_{mean} = \frac{Q}{2H} = \frac{\Delta P H^2}{3\eta L}. \tag{59}$$

If we substitute $V_{max} = \frac{\Delta P H^2}{2\eta L}$, we can again express Q and V_{mean} in terms of V_{max} as

$$Q = \left(\frac{4H}{3}\right) V_{max} \tag{60}$$

$$V_{mean} = \frac{2}{3} V_{max}. \tag{61}$$

The question is how do we locate where the average velocity V_{mean} occurs? Let us denote this location by y^*. From (45), V_{mean} at $y = y^*$ can be expressed as

$$V_{mean} = V(y^*) = \frac{\Delta P}{2\eta L} \left(H^2 - y^{*2}\right). \tag{62}$$

Comparing (62) and (61) and replacing V_{max} by (50), we obtain

$$\frac{\Delta P}{2\eta L} \left(H^2 - y^{*2}\right) = \frac{2}{3} \frac{\Delta P H^2}{2\eta L} \tag{63}$$

Table 2 Comparison of Poiseuille equations for tube and channel

Tube flow Hagen–Poiseuille	Channel flow plane Poiseuille
Fully developed laminar; steady;	Fully developed laminar; steady; uniform
Uniform flow inside tube	Flow between the two fixed plates
$V(r) = \frac{\Delta P}{4\eta L}\left(R^2 - r^2\right)$	$V(y) = \frac{\Delta P}{2\eta L}\left(H^2 - y^2\right)$
$V(r) = V_{max}\left(1 - \frac{r^2}{R^2}\right)$	$V(y) = V_{max}\left(1 - \frac{y^2}{H^2}\right)$
$V_{max} = \frac{\Delta P\, R^2}{4\eta L}$	$V_{max} = \frac{\Delta P\, H^2}{2\eta L}$
$V_{mean} = \frac{\Delta P\, R^2}{8\eta L}$	$V_{mean} = \frac{\Delta P H^2}{3\eta L}$
$V_{mean} = \frac{V_{max}}{2}$	$V_{mean} = \frac{2}{3}V_{max}$
$V_{max} = 2\,V_{mean}$	$V_{max} = \frac{3}{2}V_{mean}$
$V_{mean} = V(r^*)$ where,	$V_{mean} = V(y^*)$ where,
$r^* = \frac{R}{\sqrt{2}}$	$y^* = \frac{H}{\sqrt{3}}$
$Q = \frac{\pi\,\Delta P\, R^4}{8\,\eta\, L}$	$Q = \frac{2\Delta P\, H^3}{3\eta L}$
$Q = \left(\frac{\pi R^2}{2}\right)V_{max}$	$Q = \left(\frac{4H}{3}\right)V_{max}$

$$H^2 - y^{*2} = \frac{2H^2}{3} \tag{64}$$

$$3H^2 - 3y^{*2} = 2H^2 \tag{65}$$

$$3y^{*2} = H^2 \tag{66}$$

$$y^* = \frac{H}{\sqrt{3}}. \tag{67}$$

Hence, V_{mean} is attained at $y^* = \frac{H}{\sqrt{3}}$, which is listed in Table 2 and depicted in Fig. 12b.

Table 2 compares the Poiseuille equations for 3D tube and 2D channel flows. The velocity profiles obtained using the formulas in Table 2, and shown in Figs. 13 and 14 for tube and channel, differ by a factor of $1/2$.

We observe that the exact solution for the channel flow given by the formula (45) is independent of x. So, at every cross section of the channel, we expect to get parabolic velocity profile. Therefore, along the horizontal line, the parabolic solution is repeated for each x as shown in Fig. 15. The motion depends only on y, which gives rise to a **one-dimensional (1D) problem**. Hence, instead of solving the two-dimensional channel flow, we will describe below the numerical solution of (35) as a one-dimensional problem along the vertical axis y.

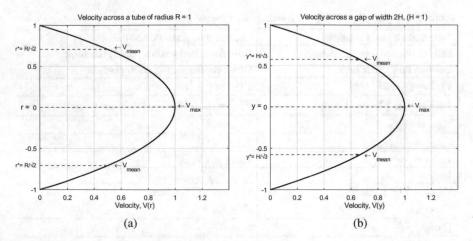

Fig. 12 Locations of V_{max} and V_{mean} for tube (**a**) and 2D channel (**b**)

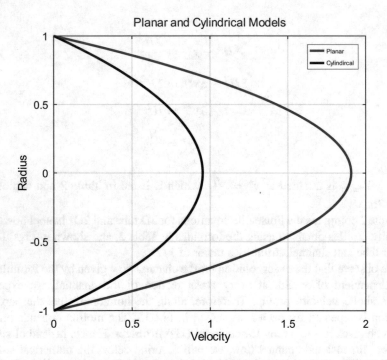

Fig. 13 Parabolic velocity profiles in tube and channel. Table 2 compares the coefficients of V_{max}

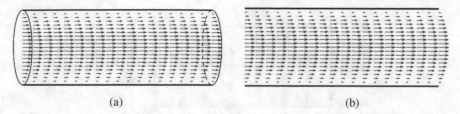

(a)

(b)

Fig. 14 Velocity plot in tube and parallel channel. (**a**) Velocity plot in a tube. (**b**) Velocity plot between two parallel plates

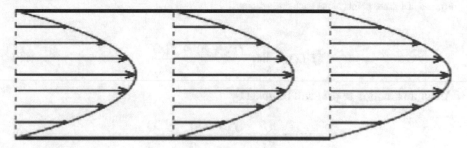

Fig. 15 Parabolas at selected cross sections in a 2D channel

4 Numerical Solution of the 1D Model

In this section, we will describe the numerical solution or the approximate solution of (35). Let us rewrite the equation here

$$\frac{d^2V}{dy^2} = -\frac{\Delta P}{\eta L}.$$

Our goal is to find the numerical solution $V(y)$ of the ordinary differential equation. To implement the numerical simulation, we need to discretize the channel as shown in Fig. 16. However, we have seen from (45) that the velocity along any cross section has a **parabolic profile**, and therefore, we do not need to discretize the entire channel to solve the ODE. Our goal here is to find the solution along any cross section and repeat it for the entire domain. For simplicity, we choose the cross section of the flow channel at $x = 0.5L$ as shown in Fig. 16. From computational point of view, this is a great advantage to understand the flow behavior along the one-dimensional vertical line. In order to achieve that, we divide the line into equal parts and evaluate the function at those discrete points. In turn, this will give the solution of the second-order differential equation (35). Observe that this Eq. (35) has a second derivative on the left-hand side, which we need to approximate. From Calculus, we know that the derivative of a function $f(x)$ is given by the following limit:

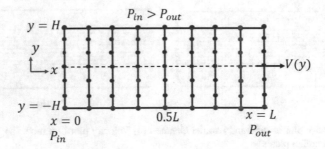

Fig. 16 Discrete points along each cross section

$$f'(x) = \lim_{h \to 0} \frac{f(x+h) - f(x)}{h}. \tag{68}$$

An approximation to (68) will be given by

$$f'(x) = \frac{df}{dx} \approx \frac{f(x+h) - f(x)}{h}. \tag{69}$$

This is also known as **forward difference formula**. Furthermore, the second derivative will be an approximation to

$$f''(x) = \frac{d^2 f}{dx^2} \approx \frac{\frac{f(x+h)-f(x)}{h} - \frac{f(x)-f(x-h)}{h}}{h} = \frac{f(x+h) - 2f(x) + f(x-h)}{h^2}. \tag{70}$$

The above formula suggests that, in order to approximate the value of the second derivative of a function at a current point x, we need to evaluate the function at three points at equal distance. These three points are the current point x and the past and future points represented by $x - h$ and $x + h$, respectively. This is also known as **central difference formula**. We will use (70) as an approximation to second derivative in order to solve the second-order differential equation (35).

Let us discretize our channel along the vertical line at $x = 0.5L$. We will consider $(N + 1)$ discrete points y_j, $j = 0\,1, 2, 3, \cdots, N$, as shown in Table 3, generated with constant interval width $\Delta y = (2H/N)$.

We know that $V(0) = 0$ and $V(y_N) = 0$ because the channel wall is stationary. This leaves us with the $(N - 1)$ interior points, $y_1, y_2, y_3, \cdots, y_{N-1}$. The numerical solution of (35) will be obtained by using **finite difference method** where the second-order derivative at each discrete point y_j is replaced with the approximation given by

$$\frac{d^2 V}{dy^2} = \frac{V_{j+1} - 2V_j + V_{j-1}}{(\Delta y)^2}, \quad (j = 1, 2, 3, \cdots, N-1). \tag{71}$$

Table 3 Discrete points

Discrete points	Discrete velocity	
$y_N = H$	$V_N = V(y_N)$	$y = H$
$y_{N-1} = -H + (N-1)\Delta y$	$V_{N-1} = V(y_{N-1})$	
$y_{N-2} = -H + (N-2)\Delta y$	$V_{N-2} = V(y_{N-2})$	Δy
$\vdots$	$\vdots$	
$\vdots$	$\vdots$	
$y_3 = -H + 3\Delta y$	$V_2 = V(y_3)$	
$y_2 = -H + 2\Delta y$	$V_2 = V(y_2)$	
$y_1 = -H + \Delta y$	$V_1 = V(y_1)$	
$y_0 = -H$	$V_0 = V(y_0)$	$y = -H$

The discrete approximation of the ordinary differential equation (35) is the following:

$$\frac{V_{j+1} - 2V_j + V_{j-1}}{(\Delta y)^2} = -\frac{\Delta P}{\eta L}. \tag{72}$$

Or,

$$V_{j+1} - 2V_j + V_{j-1} = f_j, \quad (j = 1, 2, 3, \cdots, N-1), \tag{73}$$

$$\text{where } f_j = f(y_j) = -\frac{\Delta P (\Delta y)^2}{\eta L}. \tag{74}$$

$$\text{In terms of } V_{max}, \quad f_j = -\frac{2V_{max} (\Delta y)^2}{H^2}. \tag{75}$$

Below, we have considered some examples to illustrate our discretization process from a differential equation to a system of algebraic equations to understand (73). In Example 8, we have taken only 5 points and developed the corresponding algebraic equations which we solved by hand and later explained how to solve by a MATLAB code and EXCEL spreadsheet. In Example 9, we extended it to more discrete points and applied MATLAB to obtain approximate solutions. In Example 10, we described how much error occurs due to the approximate solution while comparing the results with the exact solution. In all our computations, we are not considering the physical units.

Example 8 Using $H = 1$, $L = 5$, $\Delta P = 8.0$, and $\eta = 0.42$, solve the linear system (73) derived from the discrete equation for 5 discrete points, y_0, y_1, y_2, y_3, y_4, where $\Delta y = \frac{2H}{4}$ and $y_j = -H + j\Delta y$.

Compare your result with the exact solution obtained with (51).

Solution Since $H = 1$, we observe that $\Delta y = 0.5$, and $y_0 = -1$, $y_1 = -0.5$, $y_2 = 0$, $y_3 = 0.5$, $y_4 = 1$. Since $V_0 = 0$ and $V_4 = 0$, we only need to solve a linear system of three equations,

$$\text{when } j = 1: \quad V_2 - 2V_1 + V_0 = f_1$$
$$\text{when } j = 2: \quad V_3 - 2V_2 + V_1 = f_2$$
$$\text{when } j = 3: \quad V_4 - 2V_3 + V_2 = f_3$$
$$f_j = -\frac{2V_{max}}{H^2}(\Delta y)^2.$$

Using the given values, we obtain $V_{max} = \frac{\Delta P\, H^2}{2\eta L} = 1.90476$ in (75). After substituting $V_0 = 0$ and $V_4 = 0$ and multiplying through by (-1), the equations reduce to

$$2V_1 - V_2 = -f_1 \tag{76}$$

$$-V_1 + 2V_2 - V_3 = -f_2 \tag{77}$$

$$-V_2 + 2V_3 = -f_3 \tag{78}$$

$$\text{with} - f_j = \frac{(\Delta y)^2 (2V_{max})}{(H^2)} = 0.9523809524. \tag{79}$$

Solution of the Linear System by Hand observe that Eqs. (76)–(78) are a system of algebraic equations. Hence, the original differential equation reduces to a system of algebraic equations. One can solve the system by hand and find the values of V_1, V_2, and V_3 by elimination and substitution. Subtracting (78) from (76), we obtain

$$V_1 = V_3. \tag{80}$$

Substitute (80) in (77). This yields

$$-2V_1 + 2V_2 = 0.9523809524. \tag{81}$$

Table 4 Velocity at discrete points obtained by hand and compared with the exact solution

y_j	Velocity V_j	Exact Velocity $V(y_j)$
−1.0	0	0
−0.5	1.428571429	1.428571429
0.0	1.904761905	1.904761905
0.5	1.428571429	1.428571429
1.0	0	0

Add (81) and (78). This gives

$$V_2 = 2 \times 0.9523809524 = 1.904761905. \tag{82}$$

Now, from (76), we obtain

$$V_1 = \frac{3}{2} \times 0.9523809524 = 1.428571429. \tag{83}$$

Hence,

$$V_3 = 1.428571429. \tag{84}$$

Therefore, we obtain the velocity values at discrete points as shown in Table 4 that can be compared with the exact solution using (51).

Solution of the Linear System Using MATLAB here, we will describe how to solve the system (76)–(78) using a code. First, we observe that we can rewrite these in the matrix form. This leads to the linear system,

$$\begin{bmatrix} 2 & -1 & 0 \\ -1 & 2 & -1 \\ 0 & -1 & 2 \end{bmatrix} \begin{pmatrix} V_1 \\ V_2 \\ V_3 \end{pmatrix} = \begin{pmatrix} -f_1 \\ -f_2 \\ -f_3 \end{pmatrix}. \tag{85}$$

We can represent the linear system of three unknowns above as

$$Ax = f \tag{86}$$

$$\text{where, } A = \begin{bmatrix} 2 & -1 & 0 \\ -1 & 2 & -1 \\ 0 & -1 & 2 \end{bmatrix} \quad ; \quad x = \begin{pmatrix} V_1 \\ V_2 \\ V_3 \end{pmatrix} \quad ; \quad f = \begin{pmatrix} -f_1 \\ -f_2 \\ -f_3 \end{pmatrix}. \tag{87}$$

Solution of linear system (85) using MATLAB or GNU OCTAVE *backslash function*:

$$x = f \backslash A$$

```
% LinearSystemSolve.m %MATLAB M-file
n = 3;  % n interior points

N = n+2; % includes the two ends points

H = 1;   % Radius of the channel

L = 5;  % Length of the channel

DeltaP = 8.0;  % Pressure drop
nu = 0.42;% Viscosity
Vmax =(DeltaP*H^2)/(2*nu*L); % Maximum value of V(y)

A = [2, -1, 0; -1, 2, -1; 0, -1, 2];

dy=(2*H)/(n+1);

f(1:n)=(dy^2)*(2*Vmax)/(H^2); % In terms of Vmax
% Or,
f(1:n)=(dy^2)*DeltaP/(nu*L); % In terms of pressure

for i = 1:n
yj(i) =-H+i*dy;
V_exact(i)=Vmax*(1-(yj(i)/H)^2); % Exact solution
end

sol= A\f;     %MATLAB backslash '\' command

finalsol= [yj  sol  V_exact];

disp(finalsol)   % Solution for only interior points

%--------Print of Solution-------------
%H=1, L=5, DeltaP=8, nu=0.42, V(-1)=0, V(1)=0

   yj       sol       V_(exact)
  -1.0    0.00000      0.00000
  -0.5    1.42857      1.42857
```

0.0	1.90476	1.90476
0.5	1.42857	1.42857
1.0	0.00000	0.00000

Solution of the Linear System Using EXCEL

In this approach, we will use EXCEL array functions `'MINVERSE()'` and `'MMULT()'` together in the form `'=MMULT(MINVERSE())'` to solve the linear system of three unknowns (85).

1. Select a label row (say Row 1): enter the labels for the unknowns and the right hand side f on columns A, B, C, D.
2. (Array A): enter the elements of arrays A and f in cells A2 to C4, **cell range A2:C4**.
3. (Array f): enter the elements of arrays A and f in cells D2 to D4, **cell range D2:D4**.
4. (Solution array sol): select a range of cells equal to the number of variables. In this example, since there are 3 variables, we can select **cell range E2:E4**.
5. While the solution cells are selected, type the following expression:

   `'=MMULT(MINVERSE(A2:C4),D2:D4)'`;

   Keep the cells selected and press Ctrl+Shift+Enter. The solution will display in the solution cell range.

Example 9 Extension of Example 8. Using $H = 1$, $L = 5$, $\Delta P = 8.0$, and $\eta = 0.42$, solve the linear system (73) derived from the discrete equation for 5, 7, 9, 11, and 21 discrete points, and compare your results with the exact solution obtained with (51).

Solution Using the MATLAB code, we have displayed the solution in Tables 5, 6, 7, 8, and 9 for 5, 7, 9, 11, and 21 discrete points and compared results with the exact solution. Figure 17 shows the velocity profile for 5, 7, 11, and 21 discrete points. We observe that as we increase the number of discrete points, the velocity profile gets better and smoother.

Remark To construct the matrix A, use "spdiags" in the MATLAB code for N discrete points when N is large. The complete code is given in the Appendix.

In the following example, we would like to estimate the error that we made for considering the approximate solution obtained due to numerical simulation compared to the actual solution.

Table 5 Velocity at five discrete points

5 Discrete points			
$H = 1, L = 5, \Delta P = 8$, Viscosity $= 0.42$			
k	y_k	V_k	Exact solution (46)
0	−1.0000	0.0000000000000000	0.0000000000000000
1	−0.5000	1.4285714285714284	1.4285714285714284
2	0.0000	1.9047619047619044	1.9047619047619047
3	0.5000	1.4285714285714284	1.4285714285714284
4	1.0000	0.0000000000000000	0.0000000000000000

Table 6 Velocity at seven discrete points

7 Discrete points			
$H = 1, L = 5, \Delta P = 8$, Viscosity $= 0.42$			
k	y_k	V_k	Exact solution (46)
0	−1.0000	0.0000000000000000	0.0000000000000000
1	−0.6667	1.0582010582010579	1.0582010582010579
2	−0.3333	1.6931216931216926	1.6931216931216930
3	0.0000	1.9047619047619042	1.9047619047619047
4	0.3333	1.6931216931216930	1.6931216931216932
5	0.6667	1.0582010582010581	1.0582010582010586
6	1.0000	0.0000000000000000	0.0000000000000000

Table 7 Velocity at nine discrete points

9 Discrete points			
$H = 1, L = 5, \Delta P = 8$, Viscosity $= 0.42$			
k	y_k	V_k	Exact solution (46)
0	−1.0000	0.0000000000000000	0.0000000000000000
1	−0.7500	0.8333333333333333	0.8333333333333333
2	−0.5000	1.4285714285714284	1.4285714285714284
3	−0.2500	1.7857142857142856	1.7857142857142856
4	0.0000	1.9047619047619049	1.9047619047619047
5	0.2500	1.7857142857142860	1.7857142857142856
6	0.5000	1.4285714285714288	1.4285714285714284
7	0.7500	0.8333333333333333	0.8333333333333333
8	1.0000	0.0000000000000000	0.0000000000000000

Example 10 (Extension of Example 8) Calculate the relative error and absolute error at each of the 5 discrete points in Example 8 for the constants $H = 1, L = 5$, $\Delta P = 8$, and $\eta = 0.42$.

- The **absolute error** of a numerical solution is defined as the absolute value of the difference between the numerical solution and the exact solution.

$$\textbf{Absolute error}: \quad |\text{ Numerical solution} - \text{ Exact solution }|. \qquad (88)$$

Table 8 Velocity at eleven discrete points

11 Discrete points			
$H = 1$, $L = 5$, $\Delta P = 8$, Viscosity $= 0.42$			
k	y_k	V_k	Exact solution (46)
0	−1.0000	0.0000000000000000	0.0000000000000000
1	−0.8000	0.6857142857142857	0.6857142857142854
2	−0.6000	1.2190476190476192	1.2190476190476189
3	−0.4000	1.6000000000000003	1.6000000000000001
4	−0.2000	1.8285714285714292	1.8285714285714283
5	0.0000	1.9047619047619053	1.9047619047619047
6	0.2000	1.8285714285714292	1.8285714285714283
7	0.4000	1.6000000000000003	1.5999999999999996
8	0.6000	1.2190476190476192	1.2190476190476187
9	0.8000	0.6857142857142857	0.6857142857142854
10	1.0000	0.0000000000000000	0.0000000000000000

Table 9 Velocity at twenty-one discrete points

21 Discrete points			
$H = 1$, $L = 5$, $\Delta P = 8$, Viscosity $= 0.42$			
k	y_k	V_k	Exact solution (46)
0	−1.0000	0.0000000000000000	0.0000000000000000
1	−0.9000	0.36190476190476195	0.36190476190476178
2	−0.8000	0.68571428571428583	0.68571428571428539
3	−0.7000	0.97142857142857175	0.97142857142857142
4	−0.6000	1.2190476190476196	1.2190476190476189
5	−0.5000	1.4285714285714293	1.4285714285714284
⋮	⋮	⋮	⋮
10	0.0000	1.9047619047619055	1.9047619047619047
⋮	⋮	⋮	⋮
15	0.5000	1.4285714285714293	1.4285714285714284
16	0.6000	1.2190476190476196	1.2190476190476187
17	0.7000	0.9714285714285718	0.9714285714285710
18	0.8000	0.68571428571428583	0.68571428571428539
19	0.9000	0.36190476190476195	0.36190476190476134
20	1.0000	0.0000000000000000	0.0000000000000000

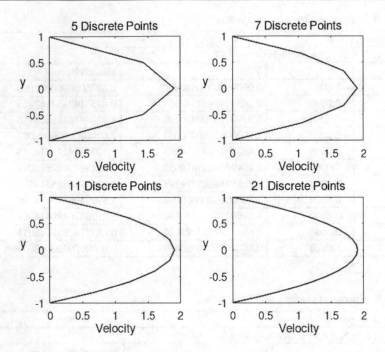

Fig. 17 Velocity profile for different discrete points. As we increase the number of discrete points, the velocity profile becomes smoother

- The **relative error** is the absolute value of the quotient whose numerator is the difference between the numerical solution and exact solution and denominator is the exact solution.

$$\textbf{Relative error} : \left| \frac{\text{Numerical solution} - \text{Exact solution}}{\text{Exact solution}} \right|. \qquad (89)$$

Solution The result is displayed in Table 10.

5   Exercises

Exercise 1 If $L = 5$, $P_1 = 10$, and $P_2 = 2$ in (30),

(a) Plot the graph of the linear function $P(x)$.
(b) Calculate the slope of the line.

Table 10 Absolute and relative errors

5 Discrete points

$H = 1, L = 5, \Delta P = 8$, Viscosity $= 0.42$

k	y_k	Numerical solution(V_k)	Exact solution (46)	Absolute error	Relative error
0	−1.0000	0.0000000000000000	0.0000000000000000	0.0000e + 00	0.0000e + 00
1	−0.5000	1.4285714285714284	1.4285714285714284	0.0000e + 00	0.0000e + 00
2	0.0000	1.9047619047619044	1.9047619047619047	2.2204e − 16	1.1657e − 16
3	0.5000	1.4285714285714284	1.4285714285714284	0.0000e + 00	0.0000e + 00
4	1.0000	0.0000000000000000	0.0000000000000000	0.0000e + 00	0.0000e + 00

Table 11 Pressure at $x = 0, 2.5,$ and 5

x	0	2.5	5
$P(x)$	10	6	2

Solution The linear function for $P(x)$ in this case is $P(x) = -\left(\frac{8}{5}\right)x + 10$. Table 11 displays the ordered pairs $(x, P(x))$, and the graph is shown in Fig. 18. The slope of the line is -1.6 which equals $\Delta P/L$.

Exercise 2 Write the equation for $V(y)$ in (46) when $H = 1, L = 5, P_1 = 10$, and $P_2 = 2$. Plot the ordered pairs $(V(y), y)$ when $\eta = 1, \eta = 2, \eta = 3$. What effect does the change in η have on the graphs?

Solution

$$\text{When } \eta = 1, \quad V(y) = 0.8\left(1 - y^2\right).$$

$$\text{When } \eta = 2, \quad V(y) = 0.4\left(1 - y^2\right).$$

$$\text{When } \eta = 3, \quad V(y) = 0.27\left(1 - y^2\right).$$

We observe that as η increases $V(y)$ decreases as shown in Fig. 19.

Exercise 3 Use the exact solution (30) to calculate the pressure, when $L = 5$, $H = 1, P_1 = 10$, and $P_2 = 2$, at the discrete points,

$$x_i = 0 + i\frac{L}{4}, \ i = 0, 1, 2, 3, 4.$$

Plot the points $(x_i, P_i), i = 0, 1, 2, 3, 4.$

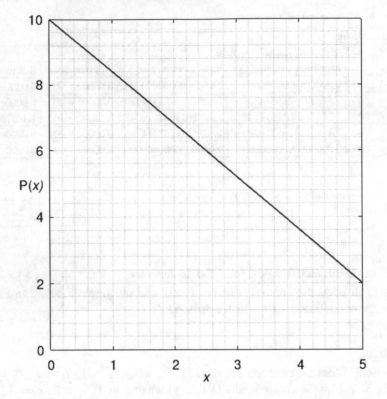

Fig. 18 Graph of $P(x)$ when $L = 5$, $P1 = 10$, and $P2 = 2$

6   Challenging Problems

Challenge Problem 1 Solve the following linear system of equations when $N = 11$ discrete points when $V_{max} = 1$:

$$V_{j+1} - 2V_j + V_{j-1} = f_j, \quad (j = 1, 2, 3, \cdots, N - 1).$$

Compare your result with the exact solution (51).

Challenge Problem 2 If the flow is in the y-direction, top to bottom, instead of the original x-direction, left to right,

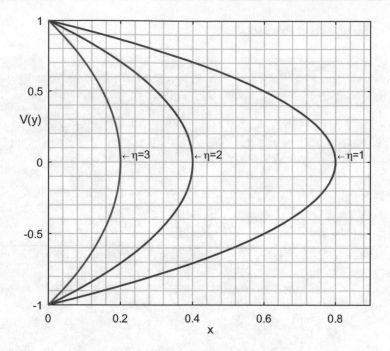

Fig. 19 Graph of $V(y)$ when $\eta = 1, 2, 3, P1 = 10$, and $P2 = 2$

(a) Write the equation for $V(x)$ in terms of x similar to (35).
(b) Write the equation for $P(y)$ in terms of y, similar to (30).
(c) Express $V(x)$ in terms of V_{max}.

Challenge Problem 3 If the flow is in the y-direction, top to bottom, instead of the original x-direction, left to right, plot the graphs of $V(x)$ and $P(y)$ from Challenge Problem 2 when $M = 5, N = 5, \eta = 1, L = 5, P_1 = 10$, and $P_2 = 2$ using the discrete points,

$$x_i = 0 + i\left(\frac{L}{M-1}\right), \quad i = 0, 1, 2, 3, \cdots, M-1.$$

$$y_j = -H + j\left(\frac{2H}{N-1}\right), \quad j = 0, 1, 2, 3, \cdots, N-1.$$

7   Suggested Projects

Research Project 1 Poiseuille Flow (Planar Model): Refer to Exercise 8 to determine the velocity at the interior points along each cross section of the channel.

Change N to 41 and use $H = 1$, $\Delta P = 8$, and $\eta = 0.42$ in (74).

Write the linear system resulting from the finite difference equation (73) for the $N - 2$ interior discrete points,

$$y_j, \; j = 1, 2, \cdots , (N - 2),$$

where $y_j = -H + j * 2H/(N - 1)$.

(a) Use MATLAB or GNU OCTAVE to solve the linear system and plot the velocity profile.
(b) Provide a table of results that compares numerical and analytic values (46) at the discrete points, and calculate at each point,
   (i)   The absolute error
   (ii) The relative error
(c) Given that $L = 5$, use MATLAB or GNU OCTAVE to plot the velocity vector over the entire length of the channel.

7.1   Cylindrical Model

Exact Solution

The planar model that we have been discussing in this chapter is an extremely simplified model. The more realistic model is the cylindrical model because it would give a better approximation of the real phenomenon. In the project that follows, we will replicate the equations for pressure and velocity in cylindrical coordinates (r, z) as shown in Fig. 20. Let us consider the cylindrical channel of radius R, $-R \leq r \leq R$, and length L. The pressure in this case is a linear function of z, and the velocity is a function of r. The governing equation (33) will be replaced by

$$\frac{\eta}{r} \frac{d}{dr} \left(r \frac{dV}{dr} \right) = \frac{dP}{dz}. \tag{90}$$

The boundary conditions of the ODE are

$$V = 0 \text{ when } r = \pm R \quad \text{and} \quad V \text{ is finite at } \quad r = 0. \tag{91}$$

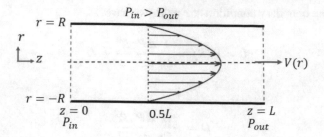

Fig. 20 Flow in a circular cylinder with radius r and length L. Flow is along the z-axis with diameter $2R$ due to pressure difference. Shown here is a horizontal slice along the diameter of the $3D$ tube to obtain a $2D$ cylindrical channel

To get the exact solution $V(r)$, we integrate (90) with respect to r to obtain

$$\int \frac{d}{dr} \left(r \frac{dV}{dr} \right) dr = \frac{1}{\eta} \frac{dP}{dz} \int r \, dr. \tag{92}$$

This will yield

$$r \frac{dV}{dr} = \frac{1}{\eta} \frac{dP}{dz} \frac{r^2}{2} + C \tag{93}$$

or

$$\frac{dV}{dr} = \frac{1}{\eta} \frac{dP}{dz} \frac{r}{2} + \frac{C}{r}, \tag{94}$$

where C is an arbitrary constant. Integrate (94) again with respect to r to get

$$\int \frac{dV}{dr} dr = \frac{1}{2\eta} \frac{dP}{dz} \int r \, dr + C \int \frac{dr}{r}. \tag{95}$$

So,

$$V(r) = \frac{1}{2\eta} \frac{dP}{dz} \frac{r^2}{2} + C \ln r + D \tag{96}$$

for arbitrary constant D. The values of C and D can be calculated as follows. We observe that when $r = 0$, at the center of the tube, velocity is finite. However, $\ln(r)$ is undefined when $r = 0$, which is a contradiction. Therefore, we assume that $C = 0$, and (96) reduces to

$$V(r) = \frac{1}{2\eta} \frac{dP}{dz} \frac{R^2}{2} + D. \tag{97}$$

Applying the boundary condition at $r = R$, we have

$$0 = V(R) = \frac{1}{2\eta} \frac{dP}{dz} \frac{R^2}{2} + D \tag{98}$$

or

$$D = -\frac{1}{2\eta} \frac{dP}{dz} \frac{R^2}{2}. \tag{99}$$

Substituting the value of D from (99) in (97), we get the exact solution. Hence, the analytic solution in cylindrical coordinates is

$$V(r) = \frac{1}{4\eta} \left(-\frac{dP}{dz}\right) \left(R^2 - r^2\right). \tag{100}$$

This equation can be written in terms of ΔP as

$$V(r) = \frac{\Delta P}{4\eta L} \left(R^2 - r^2\right). \tag{101}$$

This is the equation we described as Eq. (1) in our Introduction Sect. 1. In terms of the maximum velocity V_{max},

$$V(r) = V_{max} \left(1 - \frac{r^2}{R^2}\right). \tag{102}$$

Numerical Solution

The derivation of the finite difference approximation of the ODE (90) is a little tricky because of the division by zero at $r = 0$. Therefore, multiply (90) through by (r/η) to get

$$\frac{d}{dr} \left(r \frac{dV}{dr}\right) = \frac{r}{\eta} \frac{dP}{dz}. \tag{103}$$

Then, integrate (103) with respect to r to get

$$\int \frac{d}{dr} \left(r \frac{dV}{dr}\right) dr = \frac{1}{\eta} \frac{dP}{dz} \int r \, dr. \tag{104}$$

To derive the discrete version of the linear system of equations, we will start with the governing equation in (103) and consider small interval between two discrete

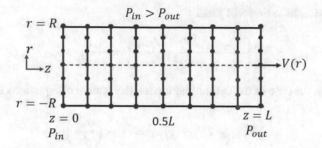

Fig. 21 Discrete points along each cross section

points r_j and r_{j+1} as shown in Fig. 21. We are considering the points along the vertical line at $z = 0.5L$. If we evaluate the integrals in (104) in the interval $[r_j, r_{j+1}]$, we get

$$\int_{r_j}^{r_{j+1}} \frac{d}{dr}\left(r\frac{dV}{dr}\right) dr = \frac{1}{\eta}\frac{dP}{dz}\int_{r_j}^{r_{j+1}} r\,dr. \tag{105}$$

This yields

$$\left[r\left(\frac{dV}{dr}\right)\right]_{r_j}^{r_{j+1}} = \frac{1}{\eta}\frac{dP}{dz}\left[\frac{r^2}{2}\right]_{r_j}^{r_{j+1}} \tag{106}$$

or

$$r\left[\left(\frac{dV}{dr}\right)_{r_{j+1}} - \left(\frac{dV}{dr}\right)_{r_j}\right] = \frac{1}{\eta}\frac{dP}{dz}\left[\frac{r^2}{2}\right]_{r_j}^{r_{j+1}}. \tag{107}$$

To derive the discrete version of the linear system of equations, we will use the approximations:

- $\Delta r = (r_{j+1} - r_j)$
- $r = \hat{r}_j = \frac{(r_j + r_{j+1})}{2}$; $\quad [r^2]_j^{j+1} = \hat{r}_j\,[r]_j^{j+1} = \hat{r}_j\,\Delta r$
- $\left(\frac{dV}{dr}\right)_j = \left(\frac{V_j - V_{j-1}}{\Delta r}\right)$
- $\left(\frac{dV}{dr}\right)_{j+1} = \left(\frac{V_{j+1} - V_j}{\Delta r}\right)$

After substituting these approximations into (107), we will obtain

$$\hat{r}_j\left[\frac{V_{j+1} - V_j - V_j + V_{j-1}}{\Delta r}\right] = \frac{dP}{dz}\hat{r}_j\frac{\Delta r}{2\eta}. \tag{108}$$

Further simplification will yield

$$\left[\frac{V_{j+1} - 2V_j + V_{j-1}}{(\Delta r)^2} \right] = \frac{1}{2\eta} \frac{dP}{dz}. \tag{109}$$

Hence, in the case of the cylindrical model, the system of equations is given by

$$V_{j+1} - 2V_j + V_{j-1} = \frac{\Delta r^2}{2\eta} \left(\frac{dP}{dz} \right) \tag{110}$$

$$V_{j+1} - 2V_j + V_{j-1} = -\Delta r^2 \left(\frac{\Delta P}{2\eta L} \right) \tag{111}$$

$$V_{j+1} - 2V_j + V_{j-1} = -\Delta r^2 \frac{2V_{max}}{R^2} \tag{112}$$

Research Project 2 Poiseuille Flow (Cylindrical Model): Refer to the finite discrete equation (112) for the cylindrical model. Given $\eta = 0.42$ and $V_{max} = 1.0$, use the discrete approximation to solve for the velocity in the cylindrical channel of radius $R = 1$ at discrete points generated with,

$$r_k = -R + k * 2R/(N - 1), \ k = 0, \ 1, \ 2, \ \cdots, \ N - 1 \ (N = 41).$$

Write the linear system resulting from the finite difference equation (112) for the $N - 2$ interior discrete points.

(a) Use MATLAB or GNU OCTAVE to solve the linear system and plot the velocity profile.
(b) Provide a table of results that compares numerical and analytic values (102) at the discrete points and calculate at each point,
  (i)   The absolute error
  (ii) The relative error
  (iii) A comparison with planar model values
 c. Given that $L = 5$, use MATLAB or GNU OCTAVE to plot the velocity vector over the entire length of the channel.

8   Conclusions and Future Work

We conclude that we have described the exact solution of a simplified 2D channel flow problem that reduces to a 1D problem. We presented the numerical solution of the 1D problem. This will help students to understand how to translate mathematical equations into applications and visualize the motion of fluid particles. This will also help students to explore how to write simple codes to model flow problems and interpret the results and how to write a project report. Additionally, this will lead to a better understanding of the applications of Calculus in finding critical values, minimum values, and maximum values of a function. Students will apply numerical methods to discretize problems and obtain approximate solutions. They can calculate errors such as absolute, relative, and percentage errors by comparing the actual results with computational approximations. We observe that velocity plays an important role in the flow problem calculations. Once maximum velocity is obtained, students can calculate other physical quantities like flow rate and mean velocity from Table 2 as these depend on velocity. By changing values of viscosity, radius, and pressure difference, students can explore different behaviors of flow pattern on their own to gain confidence. Experience gained from the scientific computing of these simple problems will lead to future exploration of modeling of more complicated and realistic biological applications.

Appendix

```
%% MATLAB M-file. (Poiseuille Flow; (x, y)-Coordinated)
% LargeN_LinearsytemY.m
% Solves [A]{u}={f}

N=21; % includes the two ends points
n=N-2;   % 19 interior points (when N=21)

H - 1;   % Radius of the channel
L = 5;   % Length of the channel

% Discrete points: y(j), j=0,1,2,3\ldots N-2,N-1.
% Interior points: y(j), j=1,2,3,---, N-2.
% y(0)=-H; y(N-1)= H.

DeltaP = 8.0;   % Pressure drop
nu = 0.42;% Viscosity
Vmax =(DeltaP*H^2)/(2*nu*L); % Maximum value of V(y)

dy=(2*H)/(n+1);
for i = 1:n
yj(i) =-H+ i*dy; %interior points
V_exact(i)=Vmax*(1-(yj(i)/H)^2); % Exact solution at
    interior points
end

% Construct matrix A using MATLAB 'spdiags' command as
    follows:
```

```
d=ones(n,1); %
A=spdiags([-1.0*d 2.0*d -1.0*d], [-1,0,1], n,n);


% f(1:n)=(dy^2)*(2*Vmax)/(H^2); % In terms of Vmax
% Or,
f(1:n)=(dy^2)*DeltaP/(nu*L); % In terms of pressure
    difference


sol= A\f;          % u=sol=V(y)


finalsol= [yj   sol   V_exact];
disp(finalsol)


% Absolute Error = |measured value - true value|
% Relative error=  |measured value - true value|/|true
    value|
```

```
%——————Print of Solution , N=21 ——————
%H=1, L=5, DeltaP=8, nu=0.42, V(-1)=0, V(1)=0
Velocity at21 Discrete Points (Poiseuille Flow; (x, y)-Coordinates)
```

yj	sol	V_(exact)	Absolute Error	Relative Error
-1.0	0.0000000000000000	0.0000000000000000	0.0000000000000000e+00	0.0000000000000000e+00
-0.9	0.3619047619047620	0.3619047619047618	1.6653345369377348e-16	4.6015822731174267e-16
-0.8	0.6857142857142858	0.6857142857142854	4.4408920985006262e-16	6.4763009769800824e-16
-0.7	0.9714285714285718	0.9714285714285714	3.3306690738754696e-16	3.4286299289894542e-16
-0.6	1.2190476190476196	1.2190476190476189	6.6613381477509392e-16	5.4643789493269433e-16
-0.5	1.4285714285714293	1.4285714285714284	8.8817841970012523e-16	6.2172489379008772e-16
-0.4	1.6000000000000005	1.6000000000000001	4.4408920985006262e-16	2.7755575615628914e-16
-0.3	1.7333333333333338	1.7333333333333334	4.4408920985006262e-16	2.5620531337503610e-16
-0.2	1.8285714285714292	1.8285714285714283	8.8817841970012523e-16	4.8572257327350609e-16
-0.1	1.8857142857142863	1.8857142857142857	6.6613381477509392e-16	3.5325278056254984e-16
0.0	1.9047619047619055	1.9047619047619047	8.8817841970012523e-16	4.6629367034256577e-16
0.1	1.8857142857142866	1.8857142857142857	8.8817841970012523e-16	4.7100370741673305e-16
0.2	1.8285714285714294	1.8285714285714283	1.1102230246251565e-15	6.0715321659188258e-16
0.3	1.7333333333333341	1.7333333333333332	8.8817841970012523e-16	5.1241062675007230e-16
0.4	1.6000000000000000	1.5999999999999996	1.1102230246251565e-15	6.9388939039072303e-16
0.5	1.4285714285714293	1.4285714285714284	8.8817841970012523e-16	6.2172489379008772e-16
0.6	1.2190476190476196	1.2190476190476187	8.8817841970012523e-16	7.2858385991025918e-16
0.7	0.9714285714285718	0.9714285714285710	7.7715611723760958e-16	8.0001365009753968e-16
0.8	0.6857142857142858	0.6857142857142854	4.4408920985006262e-16	6.4763009769800824e-16
0.9	0.3619047619047620	0.3619047619047613	6.1062266354383610e-16	1.6872468334763918e-15
1.0	0.0000000000000000	0.0000000000000000	0.0000000000000000e+00	0.0000000000000000e+00

```
%% MATLAB M-file. (Hagen-Poiseuille Flow; (r, z)-
    Coordinated)
% LargeN_LinearsytemR.m
% Solves [A]{v}={g}


N=21; % includes the two ends points
n=N-2;  % 19 interior points (when N=21)


R = 1;    % Radius of the channel
L = 5;  % Length of the channel


% Discrete points: r(j), j=0,1,2,3\ldots N-2,N-1.
% Interior points: r(j), j=1,2,3,---, N-2.
```

```
%  r(0)=-R;  r(N-1)=R.

DeltaP = 8.0;   % Pressure drop
nu = 0.42;% Viscosity
Vzmax =(DeltaP*R^2)/(4*nu*L);  % Maximum value of V(r)

dr=(2*R)/(n+1);
for i = 1:n
rj(i) =-R+ i*dr; %interior points
V_exact(i)=Vzmax*(1 - (rj(i)/R)^2); % Exact solution at
     interior points
end

% Construct matrix A using MATLAB 'spdiags' command as
     follows:
d=ones(n,1); %
A=spdiags([-1.0*d 2.0*d -1.0*d], [-1,0,1], n,n);

g(1:n)=(dr^2)*(2*Vzmax)/(R^2); %In terms of Vzmax
% Or
% g(1:n)=(dr^2)*DeltaP/(2*nu*L); %in terms of pressure
     difference

sol= A\g;          %v=sol=V(r)

finalsol= [rj   sol   V_exact];
disp(finalsol)
```

```
%————Print of Solution , N=21 —————————
%R=1, L=5,  DeltaP=8,  nu=0.42,  V(-1)=0,  V(1)=0
Velocity at 21 Discrete Points (Hagen—Poiseuille Flow; (r, z)—Coordinates)
   rj            sol                    V_(exact)                   Absolute Error            Relative Error
 -1.0   0.0000000000000000      0.0000000000000000      0.000000000000000e+00   0.0000000000000000e+00
 -0.9   0.18095238095238098     0.18095238095238089     8.3266726846886741e-17  4.6015822731174267e-16
 -0.8   0.34285714285714292     0.34285714285714269     2.2204460492503131e-16  6.4763009769800824e-16
 -0.7   0.48571428571428588     0.48571428571428571     1.6653345369377348e-16  3.4286299289894542e-16
 -0.6   0.60952380952380980     0.60952380952380947     3.3306690738754696e-16  5.4643789493269433e-16
 -0.5   0.71428571428571463     0.71428571428571419     4.4408920985006262e-16  6.2172489379008772e-16
 -0.4   0.80000000000000027     8.00000000000000004     2.2204460492503131e-16  2.7755575615628914e-16
 -0.3   0.86666666666666692     0.86666666666666670     2.2204460492503131e-16  2.5620531337503610e-16
 -0.2   0.91428571428571459     0.91428571428571415     4.4408920985006262e-16  4.8572257327350609e-16
 -0.1   0.94285714285714317     0.94285714285714284     3.3306690738754696e-16  3.5325278056254984e-16
  0.0   0.95238095238095277     0.95238095238095233     4.4408920985006262e-16  4.6629367034256577e-16
  0.1   0.94285714285714328     0.94285714285714284     4.4408920985006262e-16  4.7100370741673305e-16
  0.2   0.91428571428571470     0.91428571428571415     5.5511151231257827e-16  6.0715321659188258e-16
  0.3   0.86666666666666703     0.86666666666666659     4.4408920985006262e-16  5.1241062675007230e-16
  0.4   0.80000000000000038     0.79999999999999982     5.5511151231257827e-16  6.9388939039072303e-16
  0.5   0.71428571428571463     0.71428571428571419     4.4408920985006262e-16  6.2172489379008772e-16
  0.6   0.60952380952380980     0.60952380952380936     4.4408920985006262e-16  7.285385991025918e-16
  0.7   0.48571428571428588     0.48571428571428569     3.8857805861880479e-16  8.0001365009753968e-16
  0.8   0.34285714285714292     0.34285714285714269     2.2204460492503131e-16  6.4763009769800824e-16
  0.9   0.18095238095238098     0.18095238095238067     3.0531133177191805e-16  1.6872468334763918e-15
  1.0   0.0000000000000000      0.0000000000000000      0.0000000000000000e+00  0.0000000000000000e+00
```

Bibliography

1. Michael A. Calter, Paul A. Calter, Paul Wraight, Sarah White; *Technical Mathematics with Calculus*, John Wiley & Sons, Inc., 2011.
2. Paul Peter Urone, Roger Hinrichs; CollegePhysics, Chapter12 OpenStax, Houston Texas, 2012.
3. D. J. Acheson; *Elementary Fluid Dynamics*, Oxford University Press Inc., 1990.
4. J. D. Anderson; *Computational Fluid Dynamics: The Basics with Applications*, McGraw-Hill, Singapore.
5. J. H. Ferziger, M. Perić, R. L. Street; *Computational Methods for Fluid Dynamics*, 4ed, Springer Nature Switzerland, 2002.
6. Cleve B. Moler; *Numerical Computing with MATLAB*, SIAM, 2004.
7. Fielding H. Garrison; *An Introduction to the History of Medicine*, 3ed, W. B. Saunders Compan, Pp 495-496, 1922.
8. Andras Gedeon; *Science and Technology in Medicine* Springer, 2006.
9. Nico Westerhof, Nikos Stergiopulos, Mark I. M. Noble; *Snapshots of Hemodynamics: An aid for clinical research and graduate education*, Springer, 2005.
10. S. P. Sutera, R. Skalak; *The History of Poiseuille's Law*, Annual Review of Fluid Mechanics, Vol. 25:1-20, 1993.
11. *Poiseuille, Jean Léonard Marie*, encyclopedia.com, https://www.encyclopedia.com/science/dictionaries-thesauruses-pictures-and-press-releases/poiseuille-jean-leonard-marie.
12. Glenn Elert; *The Physics of Hypertextbook*. https://physics.info/viscosity/
13. Richard E. Klabunde; Cardiovascular Physiology Concepts. 3rd ed., Wolters Kluwer, 2021. https://www.cvphysiology.com/
14. Frank M. White; *Fluid mechanics*, 7ed, McGraw-Hill, Inc. 2009.
15. Elaine N. Marieb; *Human Anatomy & Physiology*, 6ed, Pearson, 2004.
16. James Stewart, Daniel Clegg, Saleem Watson; *Calculus*, 9ed, Cengage, 2021.
17. Wikipedia; *Hemorheology*. https://en.wikipedia.org/wiki/Hemorheology.
18. Sapna Ratan Shah; *Significance of Aspirin on Blood Flow to Prevent Blood Clotting through Inclined Multi-Stenosed Artery*. Letters in Health and Biological Sciences, 2017. https://doi.org/10.15436/2475-6245.17.018
19. Prabook *Jean Léonard Marie Poiseuille, mathematician, physicist, physiologist, scientist*. https://prabook.com/web/jean.poiseuille/3758036
20. *Unregulated artery cell growth may drive atherosclerosis, Stanford Medicine research shows*, Stanford Medicine, News Center, June 19, 2020. https://med.stanford.edu/news/all-news/2020/06/unregulated-artery-cell-growth-may-drive-atherosclerosis.html.
21. *Source*,   Source: https://www.medicinenet.com/image-collection/carotid_arteries_disease_picture/picture.htm

Statistical Tools and Techniques in Modeling Survival Data

Tom Overman and Suvra Pal

Abstract

This chapter introduces some cutting-edge tools and techniques to analyze time to event or survival data and draw associated inference. We start off by introducing different quantities of interest without which we cannot proceed with survival data analysis. We also introduce different statistical distributions, including a family of distributions, that can be very useful in modeling survival data. Since censoring is prevalent in any survival data, we show how parameter estimation can be carried out using a likelihood-based approach in the presence of right-censored data. We also provide R codes for easy implementation and believe that the readers will be motivated to develop further codes based on the specific needs of the applications. To select the best model, among a set of candidate models, that can provide a realistic description of the survival data, we present model selection methods using both likelihood ratio test and information-based criteria. Throughout the chapter, we have provided challenging problems for interested readers to solve. Finally, we have ended the chapter with three research problems that require deeper understanding of the subject area and extensive use of R software.

Suggested Prerequisites *This chapter assumes knowledge of basic probability and statistics that would be covered in an introductory statistics course. In addition, undergraduate coursework*

T. Overman
Northwestern University, Evanston, IL, USA
e-mail: tomoverman2025@u.northwestern.edu

S. Pal (✉)
University of Texas at Arlington, Arlington, TX, USA
e-mail: suvra.pal@uta.edu

© The Author(s), under exclusive license to Springer Nature Switzerland AG 2022   75
E. E. Goldwyn et al. (eds.), *Mathematics Research for the Beginning Student*,
Volume 2, Foundations for Undergraduate Research in Mathematics,
https://doi.org/10.1007/978-3-031-08564-2_3

in calculus will be helpful for many of the examples and research projects. Some experience in computer programming would be helpful but can be learned concurrently with the chapter material.

1   Introduction

Survival analysis is a branch of statistics that studies expected times of events of interest occurring. These events could be anything from the recurrence of tumors in a cancer patient to a light bulb burning out. For these reasons, survival analysis is essential in the fields of medicine, industrial engineering, and more.

The overall goal of this chapter is to introduce some tools and techniques that are commonly used in modeling all sorts of phenomena, but our examples will focus on modeling survival data. Our focus is not on a rigorous development of survival analysis, but rather an applied approach that should be enough to kick start some interesting research in modeling survival data.

There will be snippets of code throughout the text that should help the reader understand how these problems are solved computationally. These code snippets will be written in R, but the logic should be clear enough from the comments that an implementation in a different language should be entirely feasible. For a brief primer in R programming, refer to [13]. For a more detailed book on R programming with many examples of statistical methods, refer to [4]. In addition, there are a wealth of other R programming tutorials on the Internet. The code examples may be a great place to start while solving the more advanced exercises and research projects.

Our goal is to build models for survival data so that we can make predictions about a phenomenon, learn more about the process that is generating the data we are working with, and compare the viability of different models of varying complexity. In order to hop into the interesting aspects of modeling survival data, we first need to discuss some basic definitions. We will cover the probability density function, cumulative distribution function, and the survival function. We will only be working with continuous distributions in this chapter, so discrete cases will not be mentioned.

1.1   Probability Density Function

The probability density function provides the *relative* probability that a random variable takes on a particular value. It is important to note that for a continuous random variable, the absolute probability that this random variable takes a particular value is zero, but the absolute probability that the random variable is within a range of values may be greater than zero.

Definition 1 For a continuous random variable X, we can find the probability of X belonging to the range or interval $(a, b]$ with the following:

$$P(a < X \leq b) = \int_a^b f(t)dt, \tag{1}$$

where $f(\cdot)$ is the probability density function.

Exercise 1 For a continuous random variable X with probability density function $f(x) = \lambda e^{-\lambda x}$, where $x > 0$ and $\lambda > 0$, find the given probabilities in terms of λ.

(a) $P(X = 5)$
(b) $P(1 < X \leq 5)$
(c) $P(1 < X < 5)$

Exercise 2 Suppose we are hired by a light bulb manufacturing company to determine a model that predicts the probability of their light bulbs burning out at certain times after they are purchased by consumers. Using the techniques you will learn about later in this chapter, we find the data they give us to best be approximated by the following probability density function:

$$f(t) = 2.5t^{1.5}e^{-t^{2.5}},$$

where t is the number of years after purchase. This function provides the relative probability of a light bulb burning out at time t. What is the probability that the light bulb burns out within the first year using this model?

1.2 Survival Function

Now that we have some idea of the probability density function, we can define two more functions commonly used in survival analysis. The cumulative distribution function describes the probability that a random variable X will be less than or equal to a specific value x. We can formalize this in the following definition.

Definition 2 The **cumulative distribution function** for a random variable X is defined as

$$F(x) = P(X \leq x) = \int_{-\infty}^{x} f(t)dt. \tag{2}$$

Very closely related to the cumulative distribution function is the survival function. The survival function is the complement of the cumulative distribution function in that it measures the probability that a random variable X is greater than a specific value x. If our random variable T describes the time it takes for the recurrence of a disease after treatment, then the survival function $S(t)$ would describe the probability that this time to recurrence be greater than the time t. Hopefully, it is now clear why the survival function is so important for survival analysis in fields like medicine and engineering where time to failure is the defining factor.

Definition 3 We now formally define the **survival function** for a continuous random variable X with probability density function $f(x)$ as follows:

$$S(x) = P(X > x) = \int_x^\infty f(t)dt = 1 - F(x). \tag{3}$$

As you can see from the definition above, there is a fundamental relationship between the probability density function, cumulative distribution function, and survival function. Let us take a look at some examples to really dissect these new definitions.

Example 1 Given a continuous random variable X defined by probability density function $f(x) = \lambda e^{-\lambda x}$ with support $x > 0$ and $\lambda > 0$, we now wish to find the cumulative distribution function and survival function.

$$\begin{aligned} S(x) &= \int_x^\infty f(t)dt \\ &= \int_x^\infty \lambda e^{-\lambda t}dt \\ &= -e^{-\infty\lambda} + e^{-\lambda x} \\ &= e^{-\lambda x}. \end{aligned}$$

Now that we have found the survival function, we can easily find the cumulative distribution function.

$$\begin{aligned} F(x) &= 1 - S(x) \\ &= 1 - e^{-\lambda x}. \end{aligned}$$

Exercise 3 In our last example, we found the survival function and cumulative distribution function for the exponential distribution (we will discuss more on these distributions soon). Now, you are tasked with finding the survival function for the Weibull distribution. The Weibull distribution is characterized by probability density function $f(x) = \alpha\lambda x^{\alpha-1}e^{-\lambda x^\alpha}$, where $x > 0$, $\alpha > 0$, and $\lambda > 0$.

2   Common Distributions in Survival Analysis

Through some of the examples from Sect. 1, you have already been exposed to several distributions commonly encountered in survival analysis. A distribution describes the probabilities of a random variable of interest taking certain ranges of values. In this section, we will dive into some of the most important distributions used in survival analysis. It is important to note that you may see slightly altered versions of these formulas in other sources or in R functions, and you can handle this by making the appropriate change of variables. A great resource for learning more about various distributions is provided in [7].

2.1   Exponential Distribution

Definition 4 The exponential distribution has the following probability density function:

$$f(x) = \lambda e^{-\lambda x}, \tag{4}$$

where $x > 0$ and $\lambda > 0$.

In the previous section, we found the survival function and cumulative distribution function. Due to this distribution having only one parameter, it is fairly limited in the variety of shapes it can take, as shown in Fig. 1. A more detailed investigation of the exponential distribution is provided in [1].

2.2   Weibull Distribution

Definition 5 The Weibull distribution has the following probability density function:

$$f(x) = \alpha \lambda x^{\alpha-1} e^{-\lambda x^{\alpha}}, \tag{5}$$

where $x > 0$, $\alpha > 0$, and $\lambda > 0$.

We left it as an exercise to find the survival function and cumulative distribution function for the Weibull distribution. This distribution has more flexibility and can manifest in more varied shapes than the exponential distribution as shown in Fig. 2. More detailed information on the Weibull distribution is provided in [11, 14].

2.3   Gamma Distribution

Definition 6 The gamma distribution has the following probability density function:

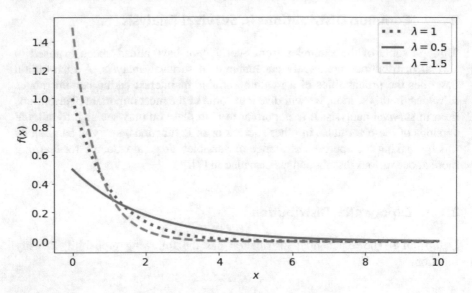

Fig. 1 Exponential probability density function with varying λ parameter values

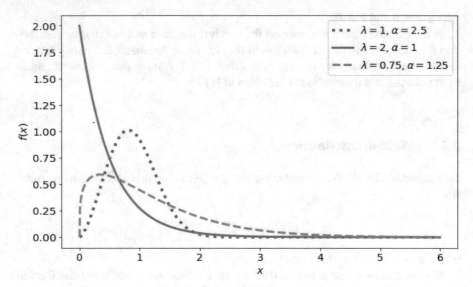

Fig. 2 Weibull probability density function with varying λ and α parameter values

$$f(x) = \frac{\lambda^\beta x^{\beta-1} e^{-\lambda x}}{\Gamma(\beta)}, \tag{6}$$

where $x > 0$, $\beta > 0$, and $\lambda > 0$. This distribution has more flexibility than the exponential distribution as shown in Fig. 3.

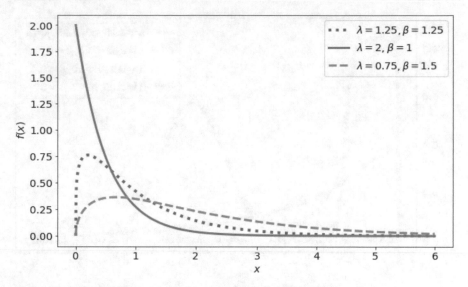

Fig. 3 Gamma probability density function with varying λ and β parameter values

You may notice the gamma function in the denominator of the gamma distribution. This gamma function is often not covered until upper level mathematics courses, so we will define it here.

Definition 7 The **gamma function** is defined for all x (except for non-positive integers) as

$$\Gamma(x) = \int_0^\infty t^{x-1} e^{-t} dt.$$

If x is a positive integer, then the formula simplifies to

$$\Gamma(x) = (x - 1)!$$

2.4 Generalized Gamma Distribution

Definition 8 The generalized gamma distribution has the following probability density function:

$$f(x) = \frac{\alpha \lambda^\beta x^{\alpha\beta-1} e^{-\lambda x^\alpha}}{\Gamma(\beta)}, \tag{7}$$

where $x > 0$, $\alpha > 0$, $\beta > 0$, and $\lambda > 0$.

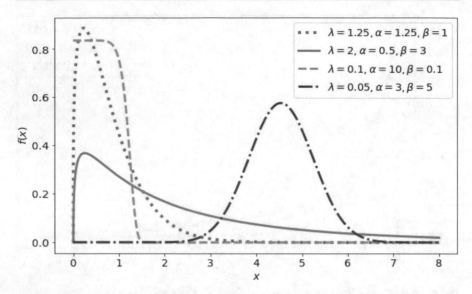

Fig. 4 Generalized gamma probability density function with varying λ, β, and α parameter values

The flexibility of the generalized gamma distribution is owed to the three different parameters used in its formulation. It should be clear from Fig. 4 that the generalized gamma has a wide variety of possible shapes, and as such can model many different interesting phenomena.

Exercise 4 As you may have noticed, the distributions have gotten increasingly complex as we have progressed through this chapter. In fact, the generalized gamma distribution can be reduced to exponential, Weibull, and gamma distributions with the right choice of parameter values. This is why we call it a *family of distributions*. See if you can find the values of α and β that reduce the generalized gamma distribution to each of its child distributions.

3   Parameter Estimation with Likelihood Function

Now that we have built up the requisite machinery to deal with distributions and their various associated functions, we are ready to start tackling a very important topic in dealing with data—parameter estimation. The idea of parameter estimation is to find the values of the parameters in our distribution that result in our distribution best matching a certain set of data. There are many ways of doing this, but we will be exploring one of the most well-known techniques called the method of maximum likelihood estimation (MLE).

3.1 Likelihood Function

The entire MLE method is dependent on what is called the likelihood function. The likelihood function is a function of our distribution's parameter values that quantifies how well our data fits the distribution with those specified parameter values.

Definition 9 Assuming we have N data points x_i that are independent and generated from the same distribution, our likelihood function is defined as

$$L(\theta) = \prod_{i=1}^{N} f(x_i),\tag{8}$$

where θ is the parameter(s) of our distribution and $f(\cdot)$ is the probability density function of our distribution.

Example 2 Let us say we have N data points and wish to find the likelihood function assuming our distribution is exponential. Then, we have

$$L(\lambda) = \prod_{i=1}^{N} \lambda e^{-\lambda x_i}.$$

As you can see from the above example, our likelihood function for the simple exponential distribution is a product of functions. In the next section, we will be maximizing the likelihood function which requires us to take derivatives. It is much easier to take derivatives if our terms are sums rather than products. For this very reason, we usually use the natural log of the likelihood function called the **log-likelihood function**.

Definition 10 The log-likelihood function is defined as

$$l(\theta) = \sum_{i=1}^{N} \ln(f(x_i)).\tag{9}$$

It is important to note that since the logarithm function is strictly increasing, the θ where $l(\theta)$ is maximized are the same θ where $L(\theta)$ is maximized. For this reason, the log-likelihood function works the same way for finding the maximum likelihood estimated parameter values but is much easier to work with.

3.2   Maximizing the Likelihood Function

Now that we have discussed the likelihood and log-likelihood functions, we are ready to tackle the rest of the MLE procedure. In order to find the parameter values that fit our data best, we need to find the value where our likelihood (or log-likelihood) function achieves a maximum.

If you have some background in calculus, you know you can use the derivative test to find the points where a function has maxima. However, for very complicated likelihood functions, performing the derivative test by hand may be entirely infeasible. For this reason, we often rely on numerical methods for finding maxima. We will discuss this more in later sections.

Example 3 We have N data points x_i that we wish to fit to an exponential distribution. Use the method of MLE to find the parameter value that best fits our data.

We start with finding a simplified expression of our log-likelihood function as follows:

$$l(\lambda) = \ln\left(\prod_{i=1}^{N} \lambda e^{-\lambda x_i}\right)$$

$$= \sum_{i=1}^{N} \ln\left(\lambda e^{-\lambda x_i}\right)$$

$$= \sum_{i=1}^{N} [\ln(\lambda) - \lambda x_i]$$

$$= N \ln(\lambda) - \lambda \sum_{i=1}^{N} x_i.$$

Now, we can use the derivative test to find the value of λ, call it $\hat{\lambda}$, that maximizes our log-likelihood function. We have

$$\frac{d}{d\lambda}\left(N \ln(\lambda) - \lambda \sum_{i=1}^{N} x_i\right) = \frac{N}{\lambda} - \sum_{i=1}^{N} x_i.$$

Setting the above equal to zero and solving for λ give our $\hat{\lambda}$ as

$$\hat{\lambda} = \frac{N}{\sum_{i=1}^{N} x_i}.$$

Note that the expression for $\hat{\lambda}$ is called the *estimator* of λ, whereas the value of $\hat{\lambda}$ obtained by plugging in the values of N and x_i is called the *estimate* of λ. Interestingly, our parameter estimate is just the reciprocal of our sample mean.

Exercise 5 Show that $\hat{\lambda}$ actually results in the maximum possible value of $l(\lambda)$. For this purpose, first take the second-order derivative of $l(\lambda)$ with respect to λ, which will be a function of λ. Then, show that this second-order derivative when evaluated at $\hat{\lambda}$ is negative.

Now, we will write some R code to maximize the log-likelihood function. While we do know the exact form of the estimate for this problem, this will generally not be the case. For this reason, we have used a built-in function to numerically find the maximum. This should help you adapt this code to more difficult problems. We use an R package called MaxLik to perform this maximization process [6]. In particular, we make use of the MaxNR function. In order for this code to work, you will need to install the MaxLik package from the CRAN repository.

```
library('maxLik')
mle_expon <- function(data){
  # set N to our sample size
  N=length(data)
  # create our log-likelihood function
  l <- function(lam){
    return(N*log(lam) - lam*sum(data))
  }
  # use MaxNR to maximize function
  estimate=maxNR(l, start = 0.0001)
# return our estimate
return(estimate)
}
```

Figure 5 shows an example of an exponential distribution approximating some data set. The parameter $\hat{\lambda}$ was determined using MLE. It should be clear that this model matches our data set quite nicely.

Exercise 6 Find the maximum likelihood estimator, $\hat{\lambda}$, for the Weibull distribution where we explicitly force $\alpha = 2$.

Challenge Problem 1 Write an R (or your programming language of choice) function that returns the maximum likelihood estimates, $\hat{\alpha}$ and $\hat{\lambda}$, for the Weibull distribution given some array of data points as an input. Later on in this chapter, you will be able to test this function with simulated data.

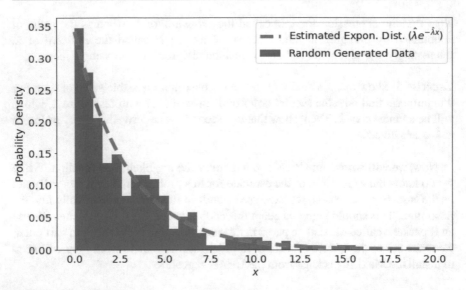

Fig. 5 MLE estimated curve for randomly generated data from exponential distribution overlaid on the histogram of the data

Hint: You will have to change the code to accommodate for maximizing two parameters. The MaxLik documentation will be helpful here. You should start by finding the proper log-likelihood function to maximize.

3.2.1 Profiling Approach for Maximizing Likelihood Functions

If you look back at the expression for the generalized gamma probability density function, you will see that there are three different parameters that we need to maximize. This makes the maximization problem much more complicated, and the numerical solver you use may not return a suitable solution. For this reason, we can use a profiling approach to overcome this problem. This approach can be used to reduce any complicated distribution with many parameters.

The main idea of the profiling approach is to reduce the complexity of the function we need to maximize by setting one or more parameter values to constants and then maximizing the resulting reduced function. Then we loop over a range of possible constant values for that parameter and repeat the maximization process on the reduced function each time while storing the result. Then, the highest value within this list of maximums is the definitive maximum, and the associated parameter values are the maximum likelihood estimates.

For example, to maximize the log-likelihood function from the generalized gamma distribution, we could loop over the values $\lambda_i = 0.01 + 0.01i$ for $i = 1, 2, 3, \ldots, 1500$ while setting λ in the log-likelihood function to that constant value. Then, we maximize this resulting two-parameter function each time, keeping track of the parameter estimates and the associated log-likelihood value. Finally, after the loop is complete, we find the highest log-likelihood value and its associated parameter estimates. For a given complicated probability distribution function, it may be challenging to determine the parameter with respect to which the profiling technique should be carried out.

3.3   Censored Data

In survival analysis, we often deal with censored data as a result of the limited nature of experimental studies. Let us say there is a cancer drug trial that measures how long it takes for a tumor to recur after successful treatment. How do we deal with cases where the tumor has not yet recurred after the end of the study? Since no study can last forever, this situation is very common in medical studies. This is an example of right-censored data, and we will briefly develop some of the machinery for dealing with this kind of data.

Definition 11 Right-Censored data occurs when a subject is removed from the study before the event of interest (e.g., recurrence of a tumor, death due to a particular disease, etc.) has taken place. In such a scenario, note that the exact time to event is not observed. However, we still have partial information on the subject, i.e., the subject survived until the time she/he was right censored. Clearly, if the subject continued the study or was followed up for a longer period of time, the exact time to event would be after the time the subject was right censored. To learn more about other types of censoring as well as truncation, refer to [2].

There are many situations where we will have to deal with right-censored data. We have some important information about right-censored data, but it would be unwise to simply treat the exit of a study as the occurrence of the event. For this reason, statisticians have devised alterations to our parameter estimation routine to properly deal with right-censored data. Although there are many other forms of censoring, we will focus mainly on right censoring as it is the most common form of censoring encountered in practice.

3.4   Parameter Estimation with Right-Censored Data

When dealing with right-censored data, we can simply alter the form of our likelihood function and carry on with the rest of the procedure as normal.

Definition 12 For right-censored data, we can incorporate our survival function into the likelihood function

$$L(\theta) = \prod_{i=1}^{N} f(x_i)^{\delta_i} S(x_i)^{1-\delta_i}, \tag{10}$$

where $\delta_i = 1$ if the ith data point is completely observed and $\delta_i = 0$ if the ith data point is right censored.

Example 4 Let us find the simplified expression of the log-likelihood function for the exponential distribution where some of our data is right censored.

$$l(\lambda) = \ln \left(\prod_{i=1}^{N} (\lambda e^{-\lambda x_i})^{\delta_i} (e^{-\lambda x_i})^{1-\delta_i} \right)$$

$$= \sum_{i=1}^{N} \ln(\lambda)^{\delta_i} - \lambda \sum_{i=1}^{N} x_i$$

$$= m \ln \lambda - \lambda \sum_{i=1}^{N} x_i,$$

where $m = \sum_{i=1}^{N} \delta_i$, the total number of completely observed data points. If we now take the first-order derivative of $l(\lambda)$ with respect to λ and then equate it to zero, we can easily show that

$$\hat{\lambda} = \frac{m}{\sum_{i=1}^{N} x_i}.$$

Note the difference between this $\hat{\lambda}$ and that obtained previously assuming all data points were completely observed.

Now, we can write some R code that will maximize this log-likelihood function based on the observed data and right-censored data.

```
library('maxLik')
mle_expon_cens <- function(obs_data, cens_data){
  m=length(obs_data)
  summed_data=sum(obs_data)+sum(cens_data)
  # create our log-likelihood function
  l <- function(lam){
    return(m*log(lam) - lam*summed_data)
  }
  # maximize the function using maxNR
  estimate = maxNR(l, start=0.0001)
  return(estimate)
}
```

Challenge Problem 2 Find the simplified expression of the log-likelihood function for the Weibull distribution that is valid for data where some of the data are right censored. Then, write an R function (or your programming language of choice) that maximizes this log-likelihood function to find the parameter estimates.

---

4   Data Simulation Methods

Before we start working with real data, we want to verify that our methods and code work properly. A common way of doing this is by simulating a data set and then testing our methods on it.

4.1   Simulation Methods for Fully Observed Data

Simulating fully observed (not censored) data is fairly straightforward due to the numerous statistical packages available in most programming languages. The goal of simulating data is to test whether our parameter estimation techniques give us valid results. For this reason, we will want to generate random data from the distribution of interest with particular parameter values. We can then measure the error between our estimate and the true parameter values.

One of the main advantages of simulating this data is that we can repeat this many times and find the average bias, average standard error, coverage probabilities, and other important quantities of interest. This allows us to directly quantify how good our estimation procedure performs.

The following R code demonstrates this simulation procedure for exponentially distributed data. It utilizes the parameter estimation function we created earlier in the chapter.

```
# n: number of random samples
# lambda: rate parameter of exponential distribution
sim_expon <- function(n, lambda){
  X = rexp(n,lambda)
  estimate = mle_expon(X)
  return(estimate)
}
```

Now, we can repeat this simulation procedure many times and calculate some important quantities relating to the estimated parameters.

```
num_sims=1000
n=100
estimates=c()
bias=c()
std_error=c()
```

```
for(i in 1:num_sims){
  #simulate with lambda=4
  simulation = sim_expon(n,4)
  estimates[i] = simulation$estimate
  hessian = simulation$hessian
  cov = -1 * solve(hessian)
  std_error[i] = sqrt(cov)
  bias[i]=estimates[i]-4
}

print(mean(estimates))
print(mean(bias))

# assuming normality of estimates, we can create
# confidence intervals for our estimates and find
# coverage probabilities
count=0
for(i in 1:num_sims){
  ci_lower=estimates[i]-std_error[i]*1.96
  ci_upper=estimates[i]+std_error[i]*1.96
  if(ci_lower<4 && ci_upper>4){
    count=count+1
  }
}
coverage_probability = count/num_sims
# we expect the coverage probability to be approx .95
print(coverage_probability)
```

4.2 Simulation Methods for Right-Censored Data

For the sake of simplicity, we will be using the term censored data to actually imply right-censored data. Simulating censored data is where things start to get more complicated. This is because we also need to assume a distribution for the censored data having a parameter representing the censoring rate, besides assuming a distribution for the complete data. A suitable value of this censoring rate needs to be calculated to achieve our desired censoring proportion for the data that we simulate. In our simulation study, we can assume the censoring distribution to be exponential with censoring rate λ_c. Let p denote the desired censoring proportion. Then, we can use the following equation to solve for the censoring rate λ_c for a given value of p:

$$p = \frac{1}{M} \sum_{i=1}^{M} S(\frac{z_i}{\lambda_c}), \tag{11}$$

where M is a large positive integer, say 10000, $S(\cdot)$ is the survival function of our distribution for the complete data, and z_i is a random value from an exponential distribution with a rate of 1. Since everything except λ_c is known in Eq. (11), we can solve this equation for λ_c. The proof for Eq. (11) is beyond the scope of this chapter, and we leave it for interested readers to come up with an understanding of the proof.

The R code below finds the censoring rate for a simulation of censored survival data where the complete data is assumed to follow an exponential distribution.

```
# n: large positive integer
# lambda: parameter of exp. dist. we are simulating
# cens_prop: desired proportion of censored data
cens_rate <- function(n, lambda, cens_prop){
  z=rexp(n,1)
  # create the function of censoring rate
  f <- function(rate){
    (1/n)*sum(exp((-lambda*z)/rate))-cens_prop
  }
  # find root of equation which is censoring rate
  censoring_rate = uniroot(f, lower=0, upper=20)
  return(censoring_rate$root)
}
```

The R code below can be used to verify if our censoring rate can accurately construct a data set with the desired censoring proportion.

```
verify_cens_rate <- function(n,lambda,cens_prop){
  # calculate censoring rate
  censoring_rate = cens_rate(10000,lambda,cens_prop)
  # create randomized data set
  X=rexp(n,lambda)
  # create censoring array
  C=rexp(n,censoring_rate)
  # create counter to track number of censored pts
  num_cens=0
  for(i in 1:n){
    # decide whether we censor the data point
    if(X[i]>C[i]){
      num_cens = num_cens + 1
    }
  }
  calc_cens_prop = num_cens/n
  return(calc_cens_prop)
}
```

Exercise 7 Write a function in R (or your programming language of choice) that calculates the censoring rate we will use to simulate right-censored data where the complete data follows a Weibull distribution. It would also be a good idea to write a code that checks if the censoring rate results in a simulated data set that has the proper censoring proportion. **Hint:** the main change you will have to make is by using the survival function of the Weibull distribution, which has two parameters.

The R code below generates the complete data set and censored data set:

```
generate_cens_data <- function(n,lambda,cens_prop){
  # create randomized data set
  X=rexp(n, lambda)
  # create censoring array assuming cens. prop.=0.4
  C=rexp(n, cens_rate(10000,lambda,0.4))
  # build out our observed and censored data sets
  cens_data=c()
  obs_data=c()
  for (i in 1:n){
    if(X[i] < C[i]){
      obs_data = c(obs_data, X[i])
    }
    else{
      cens_data = c(cens_data, C[i])
    }
  }
  return(list("cens"=cens_data, "obs"=obs_data))
}
```

The R code below generates many randomized data sets to test our estimation procedure for censored data:

```
num_sims=1000
n=100
lambda=4
cens_prop=0.4
estimates_cens=c()
bias_cens=c()
std_error_cens=c()
for(i in 1:num_sims){
  data = generate_cens_data(n,lambda,cens_prop)
  simulation = mle_expon_cens(data$obs, data$cens)
  estimates_cens[i] = simulation$estimate
  hessian = simulation$hessian
  cov = -1 * solve(hessian)
  std_error_cens[i] = sqrt(cov)
  bias_cens[i]=estimates[i]-lambda
}
```

```
print(mean(estimates_cens))
print(mean(bias_cens))

# assuming normality of estimates, we can create
# confidence intervals for our estimates and find
# coverage probability
count=0
for(i in 1:num_sims){
  ci_lower=estimates_cens[i]-std_error_cens[i]*1.96
  ci_upper=estimates_cens[i]+std_error_cens[i]*1.96
  if(ci_lower<4 && ci_upper>4){
    count=count+1
  }
}
coverage_probability_cens = count/num_sims
# we expect the coverage probability to be approx .95
print(coverage_probability_cens)
```

Challenge Problem 3 Using the code supplied as a guide, create a data simulation code to generate censored data from the Weibull distribution. You can choose the censoring proportion and Weibull parameter values. Then, test your parameter estimation code on many sets of simulated data and find some relevant quantities such as the average parameter estimates, the bias in your estimators, and the standard errors. Vary the sample size of your simulated data sets and repeat the procedure. How does the sample size affect the calculated quantities?

5   Model Selection

Now that we have learned how to estimate parameter values for any distribution, it is natural to ask which of these distributions we should use to model our data. The log-likelihood value, Akaike Information Criterion (AIC), and Bayesian Information Criterion (BIC) tests are all metrics to decide which of the models is the highest relative quality. The likelihood ratio test is a useful way to determine whether a parent distribution fits data significantly better than its constrained child distribution.

5.1   Log-Likelihood Value

The log-likelihood value is the first measure we look at after finding the parameter estimates for various distributions given a set of data. It is simply the maximum value of our log-likelihood function or, more succinctly, $l(\hat{\theta})$.

Because our likelihood function tells us how well our distribution fits our data as a function of parameter values, we can use the maximum likelihood value as a comparison tool for our various distributions. The distribution with the highest log-likelihood value fits our data best.

This is a very simple and primitive way of comparing distributions. In fact, because it does not take the number of parameters of our distribution into account, we can expect the distribution with the most parameters to have the highest log-likelihood value because it inherently has the most flexibility to fit the data. We will soon discuss ways of incorporating the complexity of the distribution into our analysis of which distribution fits our data best.

5.2   AIC and BIC

We mentioned before that just looking at the log-likelihood value is not necessarily the best way to determine which distribution is best to model our data. But why is that? Remember that we are trying to create models that work not only for one given data set but also for other data sets that could be generated from the same population. If our distribution has too many parameters, our model may be at risk of **overfitting**. An overfitted model will perform exceptionally well for the data it was built under but will have large errors for new data sets that were drawn from the same population. The danger of overfitting is illustrated in Fig. 6, where a model with many more parameters ends up severely overfitting the data. We also do not want to underfit a model and not have good prediction for any data set. For this reason, we want a measure of goodness of fit that avoids underfitting and overfitting. This is achieved by the **Akaike Information Criterion (AIC)** and **Bayesian Information Criterion (BIC)**.

Definition 13 The Akaike Information Criterion value is defined as

$$AIC = 2p - 2\ln\hat{L},$$

where p is the number of parameters in the distribution and $\hat{L} = L(\hat{\theta})$.

Definition 14 The Bayesian Information Criterion value is defined as

$$BIC = p\ln n - 2\ln\hat{L},$$

where p is the number of parameters in the distribution, $\hat{L} = L(\hat{\theta})$, and n is the sample size.

Given a set of models, the model with the smallest AIC/BIC value is the preferred model. As you can see from the equations for the AIC/BIC values, models with

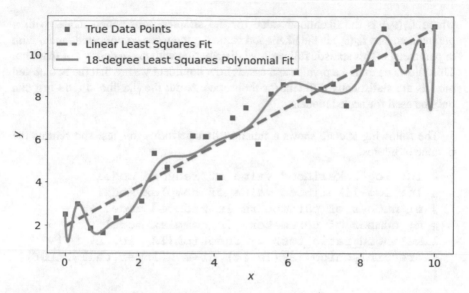

Fig. 6 Two different models for a data set. The fitted polynomial of degree 18 has many more parameters, but severely overfits the data

more parameters will have a higher value and thus are penalized for this increased complexity.

There are many other criteria that have been developed for model selection, and we encourage you to do additional research into these techniques. AIC and BIC provide simple, yet powerful, ways to select between competing models.

5.3 Likelihood Ratio Test

Throughout this chapter, we have worked with a family of distributions. All of the simpler distributions can be derived from the generalized gamma distribution with proper constraints on parameter values. Let us say we have estimated the parameters of the generalized gamma distribution and one of the simpler distributions for a particular set of data. It is natural to explore whether there is a significant difference between the generalized gamma distribution and the reduced distribution. If there is, then we can simply use the reduced distribution since it is easier to handle. We can do just that with the **Likelihood Ratio Test**.

Definition 15 The Likelihood Ratio Test statistic is defined as

$$LRT = -2 \ln \frac{L_s(\hat{\theta}_s)}{L_c(\hat{\theta}_c)} = -2[l_s(\hat{\theta}_s) - l_c(\hat{\theta}_c)],$$

where $L_s(\hat{\theta}_s)$ is the likelihood value for the simpler, constrained model with n parameters, and $L_c(\hat{\theta}_c)$ is the likelihood value for the more complicated model with m parameters. This statistic follows a χ^2 distribution with $m - n$ degrees of freedom. This allows us to find a p-value and assess how confident we are that the two nested models are statistically different (or the same). **Note:** the likelihood ratio test can only be used for nested models.

The following R code shows a function that performs this test and returns the p-value of interest.

```
# ls: log-likelihood value of reduced model
# lc: log-likelihood value of complex model
# n: number of parameters in reduced model
# m: number of parameters in complex model
likelihood_ratio_test <- function(ls, lc, n, m){
    return(pchisq(-2*(ls-lc),df=m-n,lower.tail=FALSE))
}
```

Example 5 Suppose we have a data set that we are interested in modeling under a Weibull distribution and generalized gamma distribution. We complete our estimation procedure and report the following log-likelihood values:

$$\hat{l}_{weib} = -141.1014$$

and

$$\hat{l}_{gg} = -139.9831.$$

Is the generalized gamma model significantly better at modeling the data than the Weibull model with a significance level of $\alpha = 0.05$? This requires us to use the likelihood ratio test. Our null hypothesis is that there is no statistically significant advantage in using the more complicated model. Our alternative hypothesis is that the more complicated model is significantly better at modeling our data than the reduced model. We find the p-value to be 0.1348, so it would be more appropriate to use the simpler Weibull distribution to model our data.

Exercise 8 Suppose we have a data set that we are interested in modeling under an exponential distribution and generalized gamma distribution. We complete our estimation procedure and report the following log-likelihood values:

$$\hat{l}_{exp} = -142.9345$$

and

$$\hat{l}_{gg} = -139.6266.$$

Is the generalized gamma model significantly better at modeling the data than the exponential model with a significance level of $\alpha = 0.05$?

6   Research Projects

We have discussed most of the basic machinery you will need to solve some interesting research problems in modeling survival data. The following research projects vary in difficulty, and some may require some additional research outside the scope of this chapter.

Research Project 1 This research project explores the family of distributions arising from the generalized gamma distribution. The goal is to code a parameter estimation code for each of the four distributions we have discussed (exponential, Weibull, gamma, and generalized gamma). Then, test your code with simulated data. Finally, use the real data set in Table 1 to test your code and use model selection techniques to find which model is best suited for the data. The exponential case has already been described throughout the chapter, so it can serve as a guide for the other, more complicated, distributions. The point of this project is to build some intuition with parameter estimation and other techniques we have described in the chapter.

Research Project 2 This research project closely mirrors Research Project 1 and could be used as a natural extension of the first project. However, now, the data will incorporate right-censored data. Repeat the first project, but with the right-censored data from Table 2. This will involve deriving new log-likelihood functions for each distribution, writing new parameter estimation codes, and using the more difficult data simulation methods for censored data.

Table 1 Death times for patients with cancer of the tongue [10]

Death times (weeks)
1, 3, 3, 4, 10, 13, 13, 16, 16, 24, 26, 27, 28, 30, 30, 32, 41, 51, 65, 67, 70, 72, 73, 77, 91, 93, 96, 100, 104, 157, 167

Table 2 Time to recurrence of HSV-2 [16]

Observed times to recurrence (weeks)
1, 3, 4, 5, 6, 6, 6, 8, 8, 9, 10, 12, 12, 12, 13, 14, 15, 15, 15, 19, 20, 21, 21, 27, 28, 28, 28, 35, 35, 36, 44

Censored times to recurrence (weeks)
2, 7, 9, 11, 12, 15, 24, 32, 44, 52, 52, 52, 52, 52, 52, 52, 52

Research Project 3 This research project is more open and novel than the previous two projects. It is highly advised to work on the previous projects before attempting to tackle this project. If you recall from earlier in the chapter, for complicated distributions with many parameters, such as the generalized gamma distribution, there may be a difficulty in maximizing the log-likelihood function. We proposed a very basic profiling approach to maximizing the log-likelihood function, but this could certainly be improved on. This research project is to develop a more robust and faster algorithm for maximizing the log-likelihood function for the generalized gamma distribution and similar complicated distributions. The article by Gomes et al. may be a good place to begin your investigation [5]. For more in-depth information on the generalized gamma distribution, refer to [9, 15].

7   Concluding Remarks

This chapter has hopefully been helpful in developing the basics of modeling survival data and providing some avenues for interesting and challenging research. While we have covered quite a lot of different topics, this only scratches the surface of survival analysis, and we encourage you to explore deeper into the field as you progress through the research projects. There is a wealth of resources available for those looking for deeper knowledge [3, 8, 12]; the resources we have cited throughout the chapter are great places to start.

Acknowledgments Both authors express their thanks to the Center for Undergraduate Research in Mathematics (CURM) mini-grant funded by NSF DMS-1722563. The authors also express their thanks to the following members of the UTA CURM team: Henry Alvarez, Angela Avila, and Summer Zeigman.

References

1. Balakrishnan, N. and Basu, A. P. (Eds.) (1995). The Exponential Distribution: Theory, Methods and Applications. The Netherlands: Gordon and Breach Publishers.
2. Cohen, A. C. (1991). Truncated and Censored Samples. New York: Marcel Dekker.

3. Cox, D. and Oakes, D. (1984). Analysis of Survival Data. London: Chapman & Hall.
4. Gardener, M. (2012). Beginning R: the statistical programming language. John Wiley & Sons.
5. Gomes, O., Combes, C., and Dussauchoy, A. (2008). Parameter estimation of the generalized gamma distribution. Mathematics and Computers in Simulation,79 (4), 955-963.
6. Henningsen, A. and Toomet, O. (2011). maxLik: A package for maximum likelihood estimation in R. Computational Statistics 26(3), 443-458.
7. Johnson, N. L., Kotz, S. and Balakrishnan, N. (1994). Continuous Univariate Distributions, Vol. 1, Second edition. New York: John Wiley & Sons.
8. Kalbfleisch, J. D. and Prentice, R. L. (2002). The Statistical Analysis of Failure Time Data, Second edition. New York: John Wiley & Sons.
9. Khodabin, M. and Ahmadabadi, A. (2010). Some properties of generalized gamma distribution. Mathematical Sciences 4, 9-28.
10. Klein, J. P. and Moeschberger, M. L. (2003). Survival analysis: techniques for censored and truncated data. Springer, New York, second edition.
11. Murthy, D. N. P., Xie, M. and Jiang, R. (2004). Weibull Models. Hoboken, New Jersey: John Wiley & Sons.
12. Pal, S., Yu, H., Loucks, Z. D., and Harris, I. M. (2020). Illustration of the flexibility of generalized gamma distribution in modeling right censored survival data: analysis of two cancer datasets. Annals of Data Science, 7 (1), 77-90.
13. Paradis, E. (2005). R for Beginners. Institut des Sciences de l'Evolution. Université Montpellier II.
14. Rinne, H. (2008). The Weibull Distribution: A Handbook. New York: CRC Press.
15. Stacy, E. W. (1962). A generalization of the gamma distribution. Annals of Mathematical Statistics 33, 1187-1192
16. Walker, G. and Shostak, J. (2010). Common Statistical Methods for Clinical Research with SAS Examples, Third Edition. SAS Institute.

So You Want to Price and Invest in Options?

Albert Cohen and Sooie-Hoe Loke

Abstract

The modern approach to quantitative finance is as deeply connected to the mathematical tools utilized as it is to the financial economic theory it is based upon. In this work, we employ concepts in linear algebra and optimization familiar to high school and college students to introduce them to areas of research and practice in the financial and insurance sectors. By focusing on the simplest asset evolution model, namely the binomial tree model, we define and price important financial instruments, specifically forward contracts as well as (European) put and call options. Once the reader is comfortable with this approach to pricing, we extend its application to optimal investment strategies via utility theory and pricing insurance premiums for non-traded underlying risks via indifference pricing. We provide examples in each case and propose further research projects that students and their mentors can embark upon using empirical data.

Suggested Prerequisites *Differential calculus, linear algebra, and basic probability*

A. Cohen (✉)
Department of Mathematics, Department of Statistics and Probability, Michigan State University, East Lansing, MI, USA
e-mail: acohen@msu.edu

S.-H. Loke
Department of Mathematics, Central Washington University, Ellensburg, WA, USA
e-mail: lokes@cwu.edu

© The Author(s), under exclusive license to Springer Nature Switzerland AG 2022
E. E. Goldwyn et al. (eds.), *Mathematics Research for the Beginning Student,*
Volume 2, Foundations for Undergraduate Research in Mathematics,
https://doi.org/10.1007/978-3-031-08564-2_4

1   Introduction

Consider a vector $\mathbf{X}$ of dimension N, where the ith component of $\mathbf{X}$ is denoted by x_i and represents the cash flow of an investment at time t_i, $i = 1, 2, \ldots, N$. Here, cash flow refers to a series of payments, which can be positive or negative. Furthermore, assume that one can invest a principal amount into a bank account that is compounded at an annual interest rate r. Then, we define the present value of this total cash flow as

$$
\mathrm{PV}(\mathbf{X}) := \frac{x_1}{(1+r)^{t_1}} + \frac{x_2}{(1+r)^{t_2}} + \cdots + \frac{x_N}{(1+r)^{t_N}} = \sum_{i=1}^{N} \frac{x_i}{(1+r)^{t_i}}. \tag{1}
$$

If we adopt the notation $\mathbf{v}$ for the vector whose ith element, v_i, is the present value today of a dollar paid at time t_i:

$$
v_i = \frac{1}{(1+r)^{t_i}} \tag{2}
$$

for $i = 1, 2, \ldots, N$, then we can reduce (1) to the dot product

$$
\mathrm{PV}(\mathbf{X}) = \mathbf{X} \cdot \mathbf{v}. \tag{3}
$$

In this chapter, we shall focus on the pricing of a financial instrument such as a call option on an underlying stock S *at a pre-determined time T*, which still possesses inherent cash flow risk, as the actual value of the stock S should not be known ahead of time T (absent inside information). In what follows, we will extend the notion of the cash flow vector $\mathbf{X}$ as a known sequence of payments to one that evolves sequentially, where the estimate for the $(i + 1)$th payment x_{i+1} is evolved from the observed/current value x_i using a simple binomial tree model. We point out that we start the cash flow at time t_1 and so the present value is a series that starts at $i = 1$. By using this notation, we are able to reserve x_0 for the value of the cash flow:

$$
x_0 := \mathrm{PV}(\mathbf{X}) = \sum_{i=1}^{N} \frac{x_i}{(1+r)^{t_i}}. \tag{4}
$$

1.1   Binomial Model

Let us consider the simplest model, of a stock S that takes the place of $\mathbf{X}$, with today's stock value

$$x_0 = S_0 \tag{5}$$

known from reading the newspaper today and tomorrow's value

$$x_1 = S_1 \tag{6}$$

unknown.

Let us attach a space of outcomes to S_1, denoted by the flip of a two-sided coin. If the result of our coin-tossing is heads (H), then the random variable is observed to be $S_1(H)$; otherwise, the stock is valued at $S_1(T)$ connected to an outcome of tails (T) for our coin flip. Without loss of generality, we assume that $S_1(T) < S_1(H)$. Furthermore, we assign *conditional probabilities*[1] to the events by

$$
\begin{aligned}
p &:= \mathbb{P}[S_1 = S_1(H) \mid S_0] \\
q &:= \mathbb{P}[S_1 = S_1(T) \mid S_0] = 1 - p.
\end{aligned}
\tag{7}
$$

Now, suppose that someone decides to fix at time 0 the cost of purchasing S_1, but without any inside information. Is there a way to ensure this trade with a known price (denoted by the *forward price F*)? In such a trade, agreed to at time 0 but completed at time 1,

- The initial value of this trade is 0 with no money exchanged at time 0, but
- With the exchange of the known value F for the random value S_1 at time 1.

Recall that cash flow $x_1 = S_1 - F$ is a random value from the perspective of limited information at time 0. It follows that the present value of the cash flow is zero, as no money changes hand with the initial forward contract agreement. In other words,

$$0 = \frac{x_1}{1+r} = \frac{S_1 - F}{1+r} \tag{8}$$

and so we would suggest that $F = S_1$. However, S_1 is a random variable, and F is an agreed-upon up-front price S_1, and thus is not random. In our two-state model, it follows from the reason that the locked-in forward price F should be such that

$$S_1(T) < F < S_1(H). \tag{9}$$

If this was not true, then the person either holding or selling S_1 would realize a guaranteed loss (how come?).

---

[1] For any events A and B with $\mathbb{P}(B) \neq 0$, the conditional probability $\mathbb{P}(A \mid B)$, read the probability of A given B, is given by $\frac{\mathbb{P}(A \cap B)}{\mathbb{P}(B)}$. For background information on probability, one can consult [13].

We are interested in pricing F using market information available at the time the price is agreed upon, namely the pair (S_0, r) and the probability distribution of S_1. To enable this kind of analysis, we consider a space Ω of all *paths* ω, where ω is of the form $\omega = (\omega_1, \omega_2, \ldots, \omega_N)$. This can be made quite general, but for our purposes, we shall restrict ourselves to the case where each ω_i is either a head H or tail T. With this in mind, we can extend the linear functional **present value** PV(**X**) to the **expected[2] present value**:

$$\text{EPV}(\mathbf{X}) = \sum_{i=1}^{N} \sum_{\omega=(\omega_1,\ldots,\omega_i)} \frac{x_i(\omega)}{(1+r)^{t_i}} \mathbb{P}[\omega] := \mathbb{E}\left[\sum_{i=1}^{N} \frac{x_i}{(1+r)^{t_i}} \right]. \qquad (10)$$

In this setting, we see that x_i can take any one of a multitude of possible values, each denoted by $x_i(\omega_1, \ldots, \omega_i)$ and occurring with probability $\mathbb{P}[(\omega_1, \ldots, \omega_i)]$. Note that $\text{EPV}(\mathbf{X}) = \text{PV}(\mathbf{X})$ if the coin we flip is one-sided, i.e., say $\omega_1 = \omega_2 = \cdots = \omega_N = H$ was the only possible outcome of the coin flip for any time t_i.

Now, in our specific case of the trade of a forward price F for random stock value $S_1(\omega)$, where $\omega = \omega_1 = H$ or $\omega = \omega_1 = T$, we see that (10) reduces to

$$
\begin{aligned}
0 &= \text{EPV}(\mathbf{X}) \\
&= \frac{x_1(H)}{(1+r)^1} \mathbb{P}[H] + \frac{x_1(T)}{(1+r)^1} \mathbb{P}[T] \\
&= \frac{(S_1(H) - F)}{1+r} \times p + \frac{(S_1(T) - F)}{1+r} \times (1-p) \\
&\Rightarrow F = p \times S_1(H) + (1-p) \times S_1(T) := \mathbb{E}[S_1].
\end{aligned}
\qquad (11)
$$

If we were to open the financial section of a newspaper, we would likely find quotes for forward prices, at various trade times determined by those offering such contracts. The question is what is the right value for $p = \mathbb{P}[\text{head}]$?

1.2   Brief History of Quantitative Finance: From Bachelier to Black, Scholes, and Merton

A large part of pricing event-contingent contracts depends on an accepted value for p. This is certainly the case in actuarial pricing, for example. In the financial world, the notion of *arbitrage*, where one party in a trade can make a profit with nonzero probability and zero chance of a loss, with no money up front, looms large in every trader's mind. Relying on an external party for what p should be seems attractive at first, but if this probability of an increase in an asset price is wrong,

---

[2] In general, the expected value of a discrete random variable is the sum of products of the outcome and its corresponding probability.

even by a little bit, this can lead to large losses if the volume of assets traded upon this misestimation of p is also large. So, a different model is appealing for this and other reasons. A brief history of this search is in order.

Beginning in the last century, Bachelier [1] is credited for his doctoral thesis, where he set out to analyze the Paris stock market by modeling stock *prices* as Brownian motions.[3] However, a Brownian motion can become negative in value, while a share in a limited liability corporation is non-negative in value. Fast forward to the 1970s, Merton [9], as well as Black and Scholes [2], utilized Brownian motions to model stock *returns*, which, when combined with the Itô calculus, leads to a Geometric Brownian motion for the stock values. What the Bachelier model shares with Merton, Black, and Scholes is the assumption that time is continuous, which leads to the continuous time, continuous space random walk known as the Brownian motion. For reasons that include rapid prototyping, traders and quants utilize a related model that looks at stock prices taking a discrete time random walk, where a coin is flipped at each time and the asset increases in value if a head is returned and decreases in value if a tail is returned. So, at time i, we see that the coin has been flipped i times, leading to a path ω that looks something like $\omega = (H, T, H, H, T)$, in the case where $i = 5$.

1.3 Building Tree Models

Returning to our simplest example that of a coin flip to determine the value of an asset one period of time from now, we consider the possibility of pricing similar assets over *multiple time periods*. Simply put, as in Eq. (10), the value of a contract at its inception is defined by the expected present value of all payments in the future, with the constraint that the payment value x_i at time i is dependent only on publicly available information $(\omega_1, \omega_2, \ldots, \omega_i)$.

We begin our analysis for the case $i = 1$. If we have a contract that dictates a payoff $x_1(H)$ in the case of heads and $x_1(T)$ in the case of tails, then the paradigm shift that Merton, Black, and Scholes brought to financial pricing is that the initial value x_0 of this contract is defined via *replication*[4] of the payoff, via the purchase at time 0 of Δ_0 shares of stock, with the correct amount x_0 of cash from the purchaser of the option such that the value x_1 is perfectly replicated at time 1, no more and no less.

Symbolically, this means that for $\omega_1 = H$ or T, and we are certain that

$$x_1(\omega_1) = \Delta_0 S_1(\omega_1) + (x_0 - \Delta_0 S_0)(1 + r). \tag{12}$$

---

[3] Named after the Scottish botanist Robert Brown, Brownian motion was first used to describe random motion of particles immersed in a medium (say, liquid or gas). In 1905, Albert Einstein published a seminal paper on the molecular mechanisms of Brownian motion which had subsequently revolutionized physics and provided new connections to other sciences.

[4] A replicating portfolio is a pool of assets designed to reproduce the cash flows in order to exactly match the outcomes desired written into the contract.

Again, there are two states, H and T, leading to two equations to solve for our two unknowns, Δ_0 and x_0. We can write these explicitly as

$$
\begin{aligned}
x_1(H) &= \Delta_0 S_1(H) + (x_0 - \Delta_0 S_0)(1 + r) \\
x_1(T) &= \Delta_0 S_1(T) + (x_0 - \Delta_0 S_0)(1 + r).
\end{aligned}
\tag{13}
$$

We can immediately see, by subtracting $x_1(T)$ from $x_1(H)$ in (13), that

$$
\Delta_0 = \frac{x_1(H) - x_1(T)}{S_1(H) - S_1(T)}.
\tag{14}
$$

This is incredibly helpful information for a *market maker*, who is charged with hedging (making) the instrument (contract payoff) x_1 from the underlying asset S_1. However, we still need to compute the initial price x_0, which is used by the market maker to purchase Δ_0 shares of stock S at time 0, and then invest the remaining capital $x_0 - \Delta_0 S_0$ in a safe bank account to grow at the risk-neutral rate r.

We introduce the parameters u and d, such that $d < 1 < u$ and

$$
\begin{aligned}
S_1(H) &= u S_0 \\
S_1(T) &= d S_0.
\end{aligned}
\tag{15}
$$

It follows that (13) can be written as

$$
\begin{aligned}
x_1(H) &= [x_1(H) - x_1(T)]\left(\frac{u - (1+r)}{u - d}\right) + x_0(1 + r) \\
x_1(T) &= [x_1(H) - x_1(T)]\left(\frac{d - (1+r)}{u - d}\right) + x_0(1 + r).
\end{aligned}
\tag{16}
$$

If we define the **market-derived** parameters $(\tilde{p}, \tilde{q})$:

$$
\begin{aligned}
\tilde{p} &= \frac{1 + r - d}{u - d} \\
\tilde{q} &= \frac{u - (1 + r)}{u - d} = 1 - \tilde{p},
\end{aligned}
\tag{17}
$$

then we can see that (16) can be written in terms of $\tilde{p}$ and $\tilde{q}$ as

$$
\begin{aligned}
x_1(H) &= \tilde{q}\,[x_1(H) - x_1(T)] + x_0(1 + r) \\
x_1(T) &= -\tilde{p}\,[x_1(H) - x_1(T)] + x_0(1 + r).
\end{aligned}
\tag{18}
$$

Either line in (18) leads to the price x_0 that indeed looks like an expected present value:

$$
x_0 = \frac{1}{1 + r}\left(\tilde{p} x_1(H) + \tilde{q} x_1(T)\right).
\tag{19}
$$

To ensure that (19) is in fact an expected present value, we require that $0 \leq \tilde{p} \leq 1$, which is achieved by requiring

$$dS_0 = S_1(T) < S_0(1+r) < S_1(H) = uS_0. \tag{20}$$

Financially, this means there is no reason for an investor to place the initial value S_0 of a share of stock in the market if $S_0(1+r) \geq S_1(H)$, as they could earn more in a safe investment at yearly return of r per unit of principal, and there would be no reason to invest in a safe account if the market always beats the safe rate of return, i.e., if $S_0 \leq S_1(T)$. So, we assume in our model that

$$d < 1+r < u. \tag{21}$$

The general approach to determining constraints on our market parameters that lead to an *arbitrage-free*,[5] or even *complete*,[6] market is found by employing the Fundamental Theorem of Asset Pricing (see, e.g., [11] and references therein). Let us now investigate some simple examples.

1.3.1 Example: Forward Contract

For a forward contract where $S_0 = 100$, with the time 1 stock value computed using $u = 1.3$, $d = 0.8$, and $r = 0.02$, we can calculate the risk-neutral probabilities as

$$\tilde{p} = \frac{1+r-d}{u-d} = \frac{1.02-0.80}{1.3-0.80} = 0.44 = 1 - \tilde{q}. \tag{22}$$

It follows that

$$F = \tilde{\mathbb{E}}[S_1] = \tilde{p}S_1(H) + \tilde{q}S_1(T)$$
$$= \tilde{p}uS_0 + \tilde{q}dS_0 = 0.44(1.3)(100) + 0.56(0.80)(100) = 102. \tag{23}$$

We should point out here that the initial value of an instrument that delivers one share of stock at time 1 is simply that value of the stock today, S_0. This is expressed symbolically as

$$S_0 = \frac{1}{1+r}\tilde{\mathbb{E}}[S_1]$$
$$\Rightarrow F = \tilde{\mathbb{E}}[S_1] = S_0(1+r) \tag{24}$$

---

[5] Arbitrage is the practice of buying a security in one market and selling it in another market at a higher price, profiting from this price difference without any risks. In an arbitrage-free market, the value of a security is completely based on the cash flow it generates and there is at least one way to price this product.

[6] An arbitrage-free market is complete if every security in the market has a replicating portfolio and so there is exactly one way to price this product.

and so we have a nice formula for forward prices that we can also employ for our example; for a stock worth 100 today and yearly interest rate at 0.02, we should observe that $F = (100)(1.02) = 102$, as computed above.

1.3.2 Example: Call Option

As we discussed in our historical review, the notion of a call option is highly valued by traders as it provides the option-holder the right, but not the obligation, to purchase the stock S_1 at time 1 for a user-defined (strike) price K. In terms of our cash flow analysis, the problem of pricing a call option is solving for x_0 given the payoff

$$x_1(\omega_1) = \max\{S_1(\omega_1) - K, 0\}. \tag{25}$$

Returning to our example economy in the previous subsection, where $S_0 = 100$, $u = 1.3$, $d = 0.8$, and $r = 0.02$, we return $\tilde{p} = 0.44$ and so the value of our call option is

$$
\begin{aligned}
x_0 &= \frac{1}{1+r}\tilde{\mathbb{E}}\left[\max\{S_1(\omega_1) - K, 0\}\right] \\
&= \frac{1}{1+r}\left(\tilde{p}(uS_0 - K) + \tilde{q}(0)\right) \\
&= \frac{1}{1.02}0.44(130 - K) \\
&= \frac{57.2 - 0.44K}{1.02}.
\end{aligned}
\tag{26}
$$

We note that the strike price K only makes sense between 80 and 130, as no one would buy the call if the strike price is outside of this range (why?). Also, if $K = S_0 = 100$, then $x_0 = 12.94$.

Exercise 1 (Put Option and Put–Call Parity)

- A put option gives the owner the right, but not the obligation, to sell the stock at a strike price K at maturity. Using $S_0 = 100$, $u = 1.3$, $d = 0.8$, and $r = 0.02$, show that the value of the put option is $\frac{0.56K - 44.8}{1.02}$.
- Observe that if we take the difference between the values of the call and put, we obtain

$$\frac{57.2 - 0.44K}{1.02} - \frac{0.56K - 44.8}{1.02} = \frac{102 - K}{1.02}. \tag{27}$$

Now conjecture a general formula involving the value of a call, the value of a put, discount factor, forward price, and strike price. The resulting formula is known as the put–call parity. Prove your conjecture by arguing that the combined action of buying a call and selling a put delivers one share of stock, minus K units of currency, at maturity.

1.4 Calibration of $\tilde{p}$ and $\tilde{q}$

In the previous section, we were able to price a forward contract on a stock as the risk-neutral average of the price one period in the future. Now, if we assume that our formula is correct, can we **match** the observed prices listed today on an exchange to our model? For example, if a stock is indeed 100 today, and the interest rate is in fact 0.02 for the next year, but we do not know what u and d are in the market, can we **imply** them from the observed price?

This means our estimate is

$$102 = \tilde{p}uS_0 + \tilde{q}dS_0 = 100u\frac{1+r-d}{u-d} + 100d\frac{u-1+r}{u-d} = 100(1+r) \quad (28)$$

and so there is no new information here. We should instead consider the observed option prices.

As seen in (26), we know that

$$x_0 = \frac{1}{1+r}\tilde{p}(uS_0 - K). \quad (29)$$

Now, if we have *market-observed* values for the option price x_0 for a call option contract as well as r and the initial stock value S_0, then perhaps we can **estimate** u from this empirical data. To do so, let us make some initial assumptions:

- The time to expiration can be extended to T by replacing $1 + r$ with $(1 + r)^T$, which is the value of 1 dollar left to grow with compound interest paid at yearly rate r.
- The time to expiration of the call option contract is T.
- The values of up and down multiply to 1; symbolically, we mean that $d = \frac{1}{u}$.
- Finally, the option is *at-the-money*; again, in symbols, this means that $K = S_0$.

In this setting, we have that given our empirically observed triple (x_0, S_0, r), we have the pricing formula for a T-year call option in (26), defined here as $x_0(T)$ and represented via

$$x_0(T) = \frac{1}{(1+r)^T}\left(\frac{(1+r)^T - \frac{1}{u}}{u - \frac{1}{u}}(uS_0 - S_0)\right). \quad (30)$$

This can be rewritten as

$$\frac{x_0(T)}{S_0}(1+r)^T = \left((1+r)^T u - 1\right)\frac{u-1}{u^2-1} = \frac{(1+r)^T u - 1}{u+1} \tag{31}$$

and so our **estimate** $\hat{u}$ for up, based on market observations, is

$$\hat{u} = \frac{x_0(T) + S_0(1+r)^{-T}}{S_0 - x_0(T)}. \tag{32}$$

Now, we should point out that for investors, the risk-free rate r is in fact dependent on the length of the contract. So, the risk-free rate for a 1-year contract is different and normally less than the rate for a 3-year contract, say. With this in mind, we assume that the inputs are dependent on the *term T of the contract*. So, the resulting $\hat{u}$ should also depend on T:

$$\hat{u}(T) = \frac{x_0(T) + \frac{S_0}{(1+r(T))^T}}{S_0 - x_0(T)}. \tag{33}$$

Finally, we may be tempted to use continuous rates r_T^* for compounding to allow for non-integer years T, and so using the equality

$$1 + r(T) = e^{r_T^*}, \tag{34}$$

we can write our estimate for the value of u, with a term-structure derived from the input rate r_T^* and observed market prices as

$$\hat{u}(T) = \frac{x_0(T) + S_0 e^{-r_T^* T}}{S_0 - x_0(T)}. \tag{35}$$

Research Project 1 Use market data to compute $\hat{u}$, as well as implied volatility via the Futures–Cox model.

- Using market data from any popular financial repository, compute a set of values $\hat{u}(T)$ for $T = 0.5, 1, 2, 3, 4, 5, 7, 10, 20$, and 30 for a stock of your choice. Plot these on a graph with the curve for r_T^*. Do $\hat{u}(T)$ and r_T^* jointly vary, or do they diverge for some values of T?
- Use Eq. (35) to find $\hat{u}(1)$, and use the definition of $u = e^{\sigma\sqrt{1}} = e^{\sigma}$ to *imply* an annualized volatility for the underlying stock. Explicitly, solve for the equation $\hat{u}(1) = e^{\hat{\sigma}}$.

(continued)

Table 1 Implied and observed market data for XYZ for research project 1

T	$\hat{u}(T)$	$C_0(T)$	r_T^*
0.5	?	0.59	0.0050
1	?	3.43	0.0065
2	?	4.21	0.0089
3	?	4.48	0.0100
4	?	5.01	0.0150
5	?	5.56	0.0190
7	?	8.89	0.0240
10	?	9.87	0.0500
20	?	10.98	0.0680
30	?	12.32	0.0890

- Now, compute a *historical estimate* $\bar{\sigma}$ using historical data for the stock you have chosen. Does $\hat{\sigma} = \bar{\sigma}$? If not, do they differ by a lot? Discuss with your research mentor some reasons why this may be, and see if this inequality is specific to the stock you have chosen or can be seen in other stocks that share the sector or industry as your original choice.

As an initial example of this project, consider the values for a hypothetical market with shares of a Company XYZ valued today at 100 and corresponding at-the-money call options (denoted by $C_0(T)$) as presented in Table 1. Fill in the missing $\hat{u}(T)$ values for this company. Now, repeat this exercise by gathering call option price and interest rate data for a company of your choice. A nice resource for interest data is https://www.treasury.gov/resource-center/data-chart-center/Pages/index.aspx.

2 Utility Theory

In the previous section, we determined how an initial price can be obtained for a contract that protects against future adverse stock movements. However, just as importantly, an investor may have her own opinions on how the market may evolve and may wish to put her beliefs up against the market's price structure. For example, the investor may have her beliefs on how her hometown sports team will fare in an upcoming tournament that is markedly different to how the sports betting establishment she is currently visiting has priced. How does she proceed to bet accordingly?

To develop this line of thinking further, we will need tools from both behavioral finance (utility theory) and single variable calculus.

2.1   What Is Your Level of Risk Aversion?

In models of individual consumption, more is better, but there is too much of a good thing. For example, that first bite of a potato chip is delicious, as is the second and tenth. But the hundredth? Maybe not so much. How do we capture such feelings of satisfaction when modeling human behavior as it applies to investing one's capital? There are, of course, many models that attempt to understand this phenomenon, but one of the most widely applied is that of *utility theory*. For our application, this means that a human is assigned a function that measures happiness based on the input consumption level. For example, if your asset level is x, then your happiness is measured to be $U(x)$. Now, what are the conditions we should put on U to ensure that it has some resemblance to reality?

Definition 1 A utility function is a twice differentiable function of wealth, denoted by $U(x)$, defined for $x > 0$ with the following two properties:

1. Non-satiation: utility increases with wealth. Mathematically, the first derivative $U'(x) > 0$.
2. Risk aversion (also called the law of diminishing marginal utility): marginal utility of wealth decreases as wealth increases. Here, the marginal utility is the change in utility as an additional unit is consumed. Mathematically, the second derivative $U''(x) < 0$.

Some examples of utility functions are:

1. Power utility $U(x) = \frac{1}{\lambda} x^\lambda$ for any $\lambda < 1, \lambda \neq 0$.
2. (Negative) exponential utility $U(x) = -e^{-Ax}$ for any constant $A > 0$.
3. Log utility $U(x) = \ln(x)$.

It can be easily verified that if U is a utility function, then for any constants $a > 0$ and b, the function V defined by $V(x) = aU(x) + b$ is also a utility function. In other words, utility functions are invariant under positive affine transformation. Hereafter, we will use the convention that two utility functions are equivalent when they differ only by a positive affine transformation.

2.2   Expected Utility Hypothesis

The expected utility hypothesis is a popular framework in economics and behavioral finance. For a fair game or gamble (meaning that it has an expected value of zero), the player, who may be risk-averse, can refuse to play the game under the expected utility theory. In the context of this chapter, a rational investor, when faced with a choice among a set of competing feasible investment alternatives, acts to select an investment based on the expected utility criterion whose axiomatic derivation can be found in the well-known von Neumann–Morgenstern utility theorem (see [10] for further details). The criterion asserts that the investor should select the investment that maximizes her expected utility of wealth. Moreover, if two investments yield the same expected utility, then the investor is indifferent between these two investments. The following two examples illustrate this idea.

2.2.1 Example: A Fair Game

Consider an investor with current wealth $5 and a square root utility function $U(x) = \sqrt{x}$. Assume that there is one investment available. Similar to the previous setup, a fair coin is flipped. If the result is heads, the investor wins $4 (increasing her wealth to $9). If tails, she losses $4 (decreasing her wealth to $1). The expected gain is $0.5(\$4) + 0.5(-\$4) = 0$ and so this is a fair game.

Suppose that the investor has two choices: play the game or not to play the game (i.e., do nothing). By playing the game, her expected utility is $0.5\sqrt{9} + 0.5\sqrt{1} = 2$. By doing nothing, her expected utility is $\sqrt{5} = 2.236$. Since the investor acts to maximize expected utility, she should refuse to play the game.

In this example, the expected utility of the investment is 2. The wealth value which has the same utility is $4 since $u(4) = \sqrt{4} = 2$. This value $4 is called the certainty equivalent.

Definition 2 The certainty equivalent for an investment whose outcome is given by a random variable X is a constant c such that

$$U(c) = \mathbb{E}[U(X)], \tag{36}$$

or equivalently,

$$c = U^{-1}(\mathbb{E}[U(X)]). \tag{37}$$

Note that U is an increasing function, and hence its inverse is well-defined. Thus, maximizing the expected utility is equivalent to maximizing the certainty equivalent. In general, if an investor has current wealth

- Less than c, she will consider the investment attractive.
- Greater than c, she will consider the investment unattractive.
- Exactly c, she will be indifferent between investing and doing nothing.

2.2.2 Example: Optimizing a Portfolio

Consider the power utility $U_\lambda(x) = \frac{x^\lambda - 1}{\lambda}$. If $\lambda > 1$, we have a risk-loving utility function, and if $\lambda = 1$, we get a risk-neutral utility function. For the purpose of this example, we restrict our choice of λ to be $\lambda < 1$ and $\lambda \neq 0$, meaning that the investor is risk-averse. Suppose that her current wealth is $100, and investor may choose to invest any part of her wealth in a risky asset. This risky asset returns

$$\begin{cases} -10\% \text{ with probability } 0.5, \\ +20\% \text{ with probability } 0.5. \end{cases} \tag{38}$$

The expected return of this investment is $0.5(-10\%) + 0.5(+20\%) = 5\%$. We then define α to be the proportion of wealth invested in the risky asset. That is, she invests $\$100\alpha$ and does nothing with the remaining $\$100(1-\alpha)$. Then, the two possible outcomes are:

$$\text{Bad} : 0.9(100\alpha) + 100(1-\alpha) = 100 - 10\alpha$$

$$\text{Good} : 1.2(100\alpha) + 100(1-\alpha) = 100 + 20\alpha.$$

The expected utility is

$$\begin{aligned} \mathbb{E}[U(X)] &= 0.5U(100 - 10\alpha) + 0.5U(100 + 20\alpha) \\ &= 0.5\frac{(100-10\alpha)^\lambda - 1}{\lambda} + 0.5\frac{(100+20\alpha)^\lambda - 1}{\lambda}. \end{aligned} \tag{39}$$

Next we recall that maximizing expected utility is the same as maximizing the certainty equivalent $c = U^{-1}(\mathbb{E}[U(X)])$. Here, $U_\lambda^{-1}(x) = (\lambda x + 1)^{1/\lambda}$, and subsequently the maximizer can be computed using calculus or via numerical experiments. We leave this as an exercise for the readers to verify numerically that when $\lambda = -3$, the optimal amount to invest is about $\$59.33$ and when $\lambda = -5$, the optimal amount to invest is about $\$39.21$. A precise method to compute these optimal amounts will be presented shortly. For various values of λ, the curves $U_\lambda^{-1}(x)$ are called *utility hills*. The general asset allocation portfolio optimization problem is to climb the hill (could be two or more dimensions) to find the particular portfolio at its peak.

The principle of expected utility maximization says we need to find α for which $\mathbb{E}[U(X)]$ attains its maximum value. So,

$$\begin{aligned} 0 = \frac{d}{d\alpha}\mathbb{E}[U(X)] &= \frac{d}{d\alpha}\left(\frac{0.5}{\lambda}((100-10\alpha)^\lambda + (100+20\alpha)^\lambda - 2)\right) \\ &= \frac{0.5}{\lambda}(-10\lambda(100-10\alpha)^{\lambda-1} + 20\lambda(100+20\alpha)^{\lambda-1}). \end{aligned} \tag{40}$$

Simplifying yields

$$2 = \left(\frac{100+20\alpha}{100-10\alpha}\right)^{1-\lambda}. \tag{41}$$

The positive exponent $1 - \lambda$ is called the coefficient of risk aversion and is denoted by A. We can now solve for α:

$$\alpha = \frac{100(2^{1/A} - 1)}{20 + 2^{1/A}10}. \tag{42}$$

One should check that this is indeed a maximizer. Using (42), we see that when $\lambda = -3$, $\alpha = 0.593273$ and when $\lambda = -5$, $\alpha = 0.392197$, which coincide with the earlier numerical results. As A increases (i.e., λ decreases), the investor becomes more risk-averse and as such the amount of wealth invested decreases. What about an investor with low risk aversion, say $A = 2$? It turns out that the optimal amount to invest is $121.32, which is $21.32 more than her total wealth. Suppose that the investor is able to **borrow** $21.32 without interest. Then, the optimal portfolio is to borrow the $21.32, put it together with her $100, and invest all $121.32 in the risky asset. Borrowing money to finance, a risky investment is called **leverage**. If she is unable to borrow, her optimal portfolio is to invest her entire current wealth of $100.

Exercise 2 (St. Petersburg Paradox) Consider a game of chance that consists of tossing a fair coin until a head appears. Let the random variable N be the number of trials on which the first head occurs. Consider a reward of $X = 2^N$.

- Show that the expectation of this reward does not exist.
- If this reward has utility $U(x) = \ln x$, find $\mathbb{E}(U(X))$.

Exercise 3 A decision maker has utility function $U(x) = x^{1/3}$. She is given the choice between two random amounts X_1 and X_2, in exchange for her entire present capital x_0. Suppose that

$$X_1 = \begin{cases} 8 & \text{with probability } 0.5 \\ 27 & \text{with probability } 0.5 \end{cases}$$

and

$$X_2 = \begin{cases} 1 & \text{with probability } 0.6 \\ 64 & \text{with probability } 0.4. \end{cases}$$

- Show that she prefers X_1 to X_2.
- Determine for what values of x_0 she should decline the offer.
- Give an example of an utility function in which she would prefer X_2 to X_1.

2.3 Iso-Elastic Utility Functions

So far, the utility functions in the two examples belong to a class called the iso-elastic utility functions:

$$U(x) = \begin{cases} \frac{x^\lambda - 1}{\lambda} & \text{for } \lambda < 1, \lambda \neq 0, \\ \ln x & \text{for } \lambda = 0. \end{cases} \tag{43}$$

Iso-elasticity means that if we scale the wealth by a positive constant $k > 0$, we get the "same" utility function, that is,

$$U(kx) = f(k)U(x) + g(k) \tag{44}$$

for some functions $f(k)$ and $g(k)$.

For power utility, when $\lambda \neq 0$,

$$\begin{aligned} U(kx) &= \frac{(kx)^\lambda - 1}{\lambda} \\ &= k^\lambda \frac{x^\lambda - 1}{\lambda} + \frac{k^\lambda - 1}{\lambda} \\ &:= f(k)U(x) + g(k), \end{aligned} \tag{45}$$

and for logarithmic utility (or power utility when $\lambda \to 0$),

$$U(kx) = \ln(kx) = \ln x + \ln k := U(x) + g(k). \tag{46}$$

As a consequence of the iso-elasticity property, if a given percentage of asset allocation is optimal for some current level of wealth, that same percentage asset allocation is also optimal for all other levels of wealth.

In the Optimizing a Portfolio example, with initial wealth of \$100, we found $\alpha = 0.593273$ for $\lambda = -3$ and $\alpha = 0.392197$ for $\lambda = -5$. Another iso-elastic investor with current wealth of \$500,000 (or any other wealth) will have the same optimal α's. Therefore, investors with iso-elastic utility functions have a constant attitude toward risk (expressed as a percentage of their current wealth). This property is known as **constant relative risk aversion (CRRA)**.

Next, we would like to measure how much an investor likes risk. Intuitively, the more concave a utility function is, the more it is sensitive to risk. So, a risk-averse investor will have a utility function with $-U''(x)$ large. But $-U''(x)$ is not a good measure for this purpose. Now, consider an affine transformation $V(x) = aU(x) + b$. Then, $-V''(x) = -aU''(x)$ would give a different risk aversion, even though it leads to the same decision. In order to get rid of a, we define the Pratt–Arrow absolute risk aversion function:

$$A(x) := -\frac{U''(x)}{U'(x)} = -\frac{d}{dx}\ln U'(x). \tag{47}$$

This definition implies that the utility function is completely characterize by $A(x)$. Naturally, we would assume that a wealthy person can take more risk than a poor person. This would mean that $A(x)$ should be non-increasing.

Exercise 4 Show that the risk aversion function $A(x)$ for all three utility functions (power, exponential, and logarithmic) can be written as $A(x) = (\gamma + \beta x)^{-1}$, where γ and β are constants.

Research Project 2 This project is based on a popular game-show called *Deal or No Deal*.

- To get a sense of how the game-show works, watch a random episode. For a detailed setup and analysis of the game, refer to [12].
- Discuss with your mentor some possible research questions, especially those that concern utility theory. Here are some examples:
  - If there is an opponent playing against you, such as in this game-show, how does that affect your strategy?
  - In a multi-period game like a televised game-show, does the contestant's utility remain constant, or does it evolve in measure with the audience cheering?
  - Is there an optimal wealth level where a player should cash out?

The expected utility hypothesis assumes that an investor seeks the investment which maximizes her utility function given the probabilities of the outcomes. An alternate theory is called prospect theory, which postulates that an investor makes decision based on losses and gains relative to her specific situation. Interested readers can refer to the seminal work of [8]. Another excellent yet non-mathematical reference on prospect theory is [7].

3 Maximal Utility and Indifference Pricing

In markets where there is an appetite for protection against losses in asset value, but no underlying asset in which a market maker can invest to replicate the claim, there is still room for utility-based investing to price contracts that **insure** against portfolio loss. However, we begin with an example of an investor who wishes to maximize her expected utility by investing in a risky asset. In Sect. 3.2, we extend this type of model to the case of insuring against adverse market events in an incomplete market.

3.1   Setting Up the Optimization Problem

Leveraging the tools just outlined in the previous section, we can now define the problem for an investor by the pair of utility and initial capital $(U(x), X_0)$, the space $\mathcal{X}$ of random variables that contains all possible outcomes of her portfolio, and the solution of the constrained optimization problem[7] for a subset of outcomes in $\mathcal{X}$:

$$u(X_0) := \sup_{\mathcal{A}_{X_0}} \mathbb{E}[U(X_1)] \tag{48}$$

where

$$\mathcal{A}_{X_0} := \left\{ X_1 \in \mathcal{X} \mid X_0 = \frac{1}{1+r} \tilde{\mathbb{E}}[X_1] \right\}. \tag{49}$$

3.1.1   An Initial Example of Optimal Investing

Let us consider the example of a logarithmic investor who has initial wealth $X_0 = 1000$, in a market where the safe investment yields a yearly return of 3% and a risky asset S whose initial value is $S_0 = 100$ returns time 1 values in a binomial model as

$$S_1(\omega) = \begin{cases} 150 & \omega = H \\ 90 & \omega = T. \end{cases} \tag{50}$$

However, even though we can compute the risk-neutral measure $(\tilde{p}, \tilde{q})$, the investor has her own pair of (physical) probabilities $(p, q) = \left(\frac{1}{2}, \frac{1}{2}\right)$. Explicitly, the investment problem we wish to solve can be stated as

$$u(1000) := \sup_{\mathcal{A}_{1000}} \mathbb{E}[\ln(X_1)] = \sup_{\mathcal{A}_{1000}} (pX_1(H) + qX_1(T)), \tag{51}$$

where

$$\mathcal{A}_{1000} := \left\{ X_1 \in \mathcal{X} \mid 1000 = \frac{1}{1.03} \tilde{\mathbb{E}}[X_1] \right\}$$
$$:= \left\{ (X_1(H), X_1(T)) \in \mathbb{R}^2 \mid 1000 = \frac{1}{1.03} (\tilde{p}X_1(H) + \tilde{q}X_1(T)) \right\}. \tag{52}$$

Now, how should the investor stack her bets in accordance with her beliefs (physical probability) and risk aversion (utility function) to maximize her "average happiness"? Specifically, how many shares of S should she buy and how much

---

[7] The supremum refers to the least upper bound, or the maximum, if it exists. Hence, this optimization problem is actually quite similar to the ones encountered in differential calculus, but instead of optimizing on the real line, we are optimizing over the space of random variables.

should she invest in a safe account paying 3%, to achieve $u(1000)$? By doing so, her financial advisor will have (dynamically) hedged her market investments to produce a user-based optimal return.

3.1.2 Our Initial Example of Optimal Investing as Constrained Optimization

Returning to our example, we can see that

$$\tilde{p} = \frac{(1+r) - d}{u - d} = \frac{S_0(1+r) - S_1(T)}{S_1(H) - S_1(T)} = \frac{13}{60} \tag{53}$$

and so the market is "bearish" or favorable on the outcome Tails (T). Our investor has her own belief, of course, and believes that either outcome is equally likely $p = \frac{1}{2} = q$. This means that she is much more certain of an up movement in the asset price than the market is, with $p = \frac{1}{2} > \frac{13}{60} = \tilde{p}$.

With all of this information, how many shares Δ_0 should the investor purchase to obtain her optimal outcome? We know from the section on risk-neutral pricing above that the portfolio value, depending on Δ_0, looks like

$$X_1(\omega) = \Delta_0 S_1(\omega) + (X_0 - \Delta_0 S_0)(1 + r) \tag{54}$$

and so the problem can be phrased in terms of optimizing the average expected utility *over the initial number of shares purchased*. However, we will return to our setup in (51) and (52) to phrase this problem as one of optimizing the logarithmic utility, on average, and with the constraint that variables $X_1(H) := x$ and $X_1(T) := y$ satisfying

$$1000 = \frac{1}{1.03} \left(\frac{13}{60} x + \frac{47}{60} y \right). \tag{55}$$

We complete the problem statement in these new (x, y) variables as the constrained optimization

$$u(1000) = \sup_{\left\{ (x,y) \in \mathbb{R}^2 : 1000 = \frac{1}{1.03} \left(\frac{13}{60} x + \frac{47}{60} y \right) \right\}} \left(\frac{1}{2} \ln x + \frac{1}{2} \ln y \right). \tag{56}$$

Now, our constraint can be rewritten as

$$x = \frac{(60)(1030) - 47y}{13} = \frac{61800 - 47y}{13} \tag{57}$$

and so (56) can be rewritten as

$$u(1000) = \sup_{y \in \mathbb{R}} \left(\frac{1}{2} \ln y + \frac{1}{2} \ln \left(\frac{61800 - 47y}{13} \right) \right)$$

$$= \frac{1}{2} \sup_{y \in \mathbb{R}} \left(\ln y + \ln (61800 - 47y) \right) - \frac{1}{2} \ln (13).$$

(58)

It follows that for the function $f(y) = \ln y + \ln (61800 - 47y)$, we will see a maximum of f at a critical point y_c, defined by

$$f'(y_c) = \frac{1}{y_c} - \frac{47}{61800 - 47y_c} = 0.$$

(59)

Arranging terms, we have $y_c = \frac{61800}{94}$ (check that this is indeed a maximizer). Returning to our financial interpretation, we insert our value for y_c to obtain our *optimal* outcome in the case of tails, $\hat{X}_1(T) = \frac{61800}{94}$, and *optimal* outcome in the case of heads, $\hat{X}_1(H) = \frac{61800 - 47(\frac{61800}{94})}{13} = \frac{61800}{26}$.

Recognizing that this is the optimal outcome, based on the investor's utility and belief regarding market movements, we should figure out how her investor should purchase or sell shares initially to get such an outcome. This is computed by the portfolio approach: if $\omega = H$ or $\omega = T$, the advisor should return the outcome from investing in both stocks and a bank account at time 0:

$$\hat{X}_1(\omega) = \Delta_0 S_1(\omega) + (X_0 - \Delta_0 S_0)(1 + r)$$

$$= \Delta_0 S_1(\omega) + (1000 - 100\Delta_0)(1.03).$$

(60)

This results in the pair

$$\hat{X}_1(H) = \frac{61800}{26} = 150\Delta_0 + (1000 - 100\Delta_0)(1.03) = 1030 + 47\Delta_0$$

$$\hat{X}_1(T) = \frac{61800}{94} = 90\Delta_0 + (1000 - 100\Delta_0)(1.03) = 1030 - 13\Delta_0$$

(61)

and so our hedging strategy is to purchase

$$\Delta_0 = \frac{\hat{X}_1(H) - \hat{X}_1(T)}{S_1(H) - S_1(T)} = \frac{\frac{61800}{26} - \frac{61800}{94}}{150 - 90} = 28.66$$

(62)

shares. In this case, the positive sign of Δ_0 means the advisor buys 28.66 shares, in accordance with the fact that the investor has a belief that the asset will increase in value with probability $\frac{1}{2} > \frac{13}{60}$.

Two questions arise immediately:

1. How can we extend this to a general framework?
2. How can we use the notion of utility to compute insurance premiums for adverse price movements when a market maker cannot replicate payoffs due to lack of an underlying asset?

For the first question, the answer lies within the techniques of constrained optimization and Lagrange multipliers (see Challenge Problem 1). One can expand the model to include more than two states at time 1, but this would require multivariable calculus, which is beyond the scope of this introductory chapter. We will explore the second question in the next section.

Exercise 5 Consider an economy where $(S_0, r, u, d) = (100, 0.05, \frac{3}{2}, \frac{2}{3})$. What is the optimal number Δ_0 of shares that a financial advisor should purchase to achieve an optimal outcome for an investor whose utility function $U(x) = \sqrt{x}$?

Challenge Problem 1 Extend the example in this section to a general framework with arbitrary risk-neutral probabilities $(\tilde{p}, \tilde{q})$, arbitrary physical probabilities (p, q), and optimal outcomes $(\hat{X}_1(H), \hat{X}_1(T)) = (\hat{u}x, \hat{d}x)$ where $X_0 = x$. Then, use Lagrange multiplier to prove that there exists a real number λ such that

$$u'(\hat{X}_1(H)) = \lambda \frac{\tilde{p}}{p},$$

$$u'(\hat{X}_1(T)) = \lambda \frac{1 - \tilde{p}}{1 - p}.$$

(63)

Research Project 3 Suppose that we have a power utility function $U(x) = x^\alpha$. Can we determine α based on one's $p, \tilde{p}, u$, and d?

- Explain to 10 friends the setup of the investment example in this section, and then ask them for their belief (p, q), market probabilities $(\tilde{p}, \tilde{q})$, and their acceptable returns $(\hat{u}, \hat{d})$.
- Show that (63) reduces to a single functional differential equation which does not involve λ.
- Assuming power utility, find the corresponding α for each friend. Are most of your friends risk-averse, risk-neural, or risk-seeking?
- What is the relationship between p and $\tilde{p}$ for a risk-averse investor?

3.2   An Example of Indifference Pricing

Let us return to the fundamental example of a logarithmic investor with initial capital X_0. However, in our current setting, she is looking for protection against adverse movements in her portfolio and desires to purchase "option-like" protection against possible portfolio losses. Unfortunately, the objects she wishes to protect are not very well-traded, and even though they have high market value, the market is not complete. The question is how do we price the value V_0 of her option protection? This is an excellent example of the overlap between finance and insurance, where the portfolio could include works of art or other assets that are not frequently traded on a market, or risks such as protection against weather events that, of course, cannot be replicated.

So, for our investor, we can argue that the price we charge, since it cannot be determined from a risk-neutral average, is established rather by finding V_0 where the investor is *indifferent* to the utility she derives immediately or on average (with respect to her physical measure) after possibly suffering a loss in her portfolio value (denoted by the event $\{\omega \in \Omega \mid X_0 > X_1(\omega)\}$):

$$\ln(X_0 - V_0) = \mathbb{E}\left[\ln\left(X_0 - (X_0 - X_1)_+\right)\right]. \tag{64}$$

For our investor's initial portfolio of 1000, in a painting that she owns, she is looking to buy insurance against it falling in value. Pundits have determined that in one year, the painting could be worth

$$X_1(\omega) = \begin{cases} 1500 & \omega = H \\ 900 & \omega = T, \end{cases} \tag{65}$$

and the investor believes that $p = \frac{1}{2} = q$. So, it follows that

$$\ln(1000 - V_0) = \frac{1}{2}\ln(1000 - 0) + \frac{1}{2}\ln(1000 - 100) = \ln(948.68) \tag{66}$$

$$\Rightarrow V_0 = 51.32.$$

Notice that the extra 1.32 (also called the *utility premium* of this insurance) is due to the expectation under utility. Without utility (i.e., under physical probability), the premium should be 50, not 51.32. The possible loss is capped at 100 with expected frequency of 50% and so without discounting for the time-value of money, we would expect that such insurance would cost 50. However, due to the lack of risk-neutrality in this insurance, the effect of the sublinear utility function results in a cost of 51.32 instead of the expected 50.

Exercise 6 Suppose that a client has a log utility function and carries a risk X where

$$X = \begin{cases} 0 & \text{with probability } 0.5 \\ 64 & \text{with probability } 0.5. \end{cases}$$

- Assume that the insurer covers *half* of all losses and that the capital $X_0 = 100$. What is the maximum premium that the client is willing to pay for the contract?
- Assume that the insurer's minimum premium to take over the risk is 20 and that the insurer has the same utility function. If the insurer covers *half* of all losses, determine the capital of the insurer.

Research Project 4 Let us implement the indifference pricing model to pricing risks that cannot be replicated via hedging (investment). Specifically, we use Eq. (64) and auto claim data to price car insurance.

- Using market data from any popular data repository, compute a set of 1-year premiums $\hat{p}(K, X_0)$ for claims of size $K = 0.1X_0, 0.25X_0, 0.50X_0, 0.75X_0$, and $0.90X_0$, where X_0 is the value of the car at the beginning of the year. How does the premium vary with the value of X_0? Linearly? Also, what is the utility premium here?
- Specifically, solve the equation

$$\ln(X_0 - \hat{p}(K, X_0)) = \mathbb{E}\left[\ln\left(X_0 - (X_0 - K)_+ \right) \right]$$
$$= p \ln (X_0 - 0) + (1 - p) \ln (X_0 - K), \tag{67}$$

where p is an estimate, using your dataset, of the probability of a claim being made on the vehicle within the year. Assume that claims are settled at the end of the year.
- Price the insurance using a square root utility instead:

$$\sqrt{X_0 - \hat{p}(K, X_0)} = \mathbb{E}\left[\sqrt{X_0 - (X_0 - K)_+}\right] = p\sqrt{X_0 - 0} + (1 - p)\sqrt{X_0 - K}. \tag{68}$$

---

4   Conclusion and Further Reading

In this chapter, we set out to introduce basic techniques used in pricing insurance contracts and financial instruments, employing mathematical tools familiar to college students. As students progress into multivariable calculus and probability classes, they should find multiple opportunities to extend the concepts presented in the earlier sections. In what follows, we discuss some possible generalizations.

The cash flow $\mathbf{X}$ in Sect. 1 can be extended to a vector of infinite dimension. The present value of the cash flow is then given by

$$\sum_{i=1}^{\infty} \frac{x_i}{(1+r)^{t_i}}. \tag{69}$$

In a similar fashion, the expected present value (10) can also be extended to an infinite series.

The linear functional approach in finding the present value (3) can be generalized to many environments where both the yearly interest rate r and stream of payments $\mathbf{X}$ are fully known in advance of the calculation of $PV(\mathbf{X})$. However, investors accept multiple risks when applying this assumption, such as *interest rate risk* (rates fluctuate in a non-deterministic manner) as well as *cash flow risk*, defined here to be the risk associated with assuming $\mathbf{X}$ to be a deterministic vector rather than a vector of random variables.

Multiple industries require the valuation of this series of payments represented by $\mathbf{X}$. For example, when pricing life insurance, the vector $\mathbf{X}$ may be associated with a life-contingent event at a random time that provides a single, deterministic payment (also known as an insurance benefit). There is also the case where uncertainty in both the timing and the size of the payments in $\mathbf{X}$ is intertwined, such as in hybrid products like variable annuities, but we leave this emerging area of study for another time. For an introductory exposition of the topic, interested readers can consult [5].

A solid reference for the overlap in risk management tools in finance and insurance can be found in [4]. A general reference, for both discrete and continuous time, can be found in [6].

While our suggested projects do not require heavy programming, we note that many interesting research problems in math finance call for advanced programming and statistical skills and as such we strongly encourage the readers to explore some computational models. The book [3] serves as an excellent reference, with many examples in Python, R, Perl, and MATLAB. It is our hope that the readers see the connection between a structural model and the need to calibrate that theory with empirical data, so they can further engineer those models to more closely reflect the data they observe.

Acknowledgments Albert Cohen is grateful for the training in mathematical finance he received as a graduate student at Carnegie Mellon University, especially from Steven Shreve, Dmitry Kramkov, and David Heath. We recommend [14] for a deep introduction to binomial asset pricing that has guided Albert in his study and teaching of the subject. Albert is also appreciative of the support and mentorship he has received from the mathematical, actuarial, and quantitative finance communities and would like to thank Nick Costanzino, Harvey Stein, Peter Carr, Jacob Geyer, Ron Simon, Jonathan Culbert, Sean Hilden, Martin Jones, Robert J. Rietz, Garrett Mitchener, Frederi Viens, and Keith Promislow. Finally, both Albert and Sooie-Hoe would like to offer their eternal gratitude to all of their students, past, current, and future, for their grace, patience, and dedication to learning more about the field of mathematical finance, risk management, and actuarial science.

References

1. L. Bachelier. Théorie de la spéculation. In *Annales scientifiques de l'École normale supérieure*, volume 17, pages 21–86, 1900.
2. F. Black and M. Scholes. The pricing of options and corporate liabilities. *Journal of political economy*, 81(3):637–654, 1973.
3. S. R. Dunbar. *Mathematical Modeling in Economics and Finance: Probability, Stochastic Processes, and Differential Equations*, volume 49. American Mathematical Soc., 2019.
4. P. Embrechts. Actuarial versus financial pricing of insurance. *The Journal of Risk Finance*, 2000.
5. R. Feng. *An introduction to computational risk management of equity-linked insurance*. CRC Press, 2018.
6. V. Henderson and D. Hobson. Utility indifference pricing-an overview. *Volume on Indifference Pricing*, 2004.
7. D. Kahneman. *Thinking, fast and slow*. Macmillan, 2011.
8. D. Kahneman and A. Tversky. Prospect theory: An analysis of decision under risk. In *Handbook of the fundamentals of financial decision making: Part I*, pages 99–127. World Scientific, 2013.
9. R. C. Merton. Theory of rational option pricing. *The Bell Journal of economics and management science*, pages 141–183, 1973.
10. L. J. Neumann, O. Morgenstern, et al. *Theory of games and economic behavior*, volume 60. Princeton university press Princeton, 1947.
11. A. Pascucci. *PDE and martingale methods in option pricing*. Springer Science & Business Media, 2011.
12. T. Post, M. J. Van den Assem, G. Baltussen, and R. H. Thaler. Deal or no deal? decision making under risk in a large-payoff game show. *American Economic Review*, 98(1):38–71, 2008.
13. S. M. Ross. *Introduction to probability models*. Academic press, 2014.
14. S. Shreve. *Stochastic calculus for finance I: the binomial asset pricing model*. Springer Science & Business Media, 2005.

The Spiking Neuron Model

Maxwell E. Bohling and Lawrence C. Udeigwe

Abstract

The human nervous system is constantly responding to stimuli as we interact with our surrounding environment. Information pertaining to a stimulus is encoded in electrochemical signals transmitted by specialized nerve cells called neurons. If the intensity of a stimulus reaches a certain threshold, a signal is generated and transmitted as the result of an action potential or spike. This describes the basis for communication between the central nervous system and the rest of the body. This chapter is devoted to the presentation and implementation of several spiking neuron models beginning with the integrate-and-fire model. We take the reader through a short treatment of physical structure and electrical properties of a neuron. This will provide the background needed to understand model parameters as well as how to implement them in Python and how to run simulations of their own.

Suggested Prerequisites *Familiarity with single-variable calculus, basic probability theory, and basic programming is recommended though not required.*

1   Introduction

Imagine you have been hard at work finishing up a research paper. Hours have passed and the sun has set without you noticing. That is, until your roommate inevitably interrupts, asking why you are working in the dark. Before you can

M. E. Bohling · L. C. Udeigwe (✉)
Manhattan College, Riverdale, NY, USA
e-mail: mbohling01@manhattan.edu; lawrence.udeigwe@manhattan.edu

© The Author(s), under exclusive license to Springer Nature Switzerland AG 2022   127
E. E. Goldwyn et al. (eds.), *Mathematics Research for the Beginning Student*,
Volume 2, Foundations for Undergraduate Research in Mathematics,
https://doi.org/10.1007/978-3-031-08564-2_5

answer, they flip the light switch on unexpectedly, leading to a sharp change in light intensity. Naturally, you put up a hand to shade your eyes as they adjust. This illustrates a simple example of a bodily response to a sensory *stimulus*. Our bodies are constantly responding to stimuli as we interact with our surrounding environment. If the intensity of a stimulus reaches a certain threshold value, neurons are *activated*, that is, they begin to transmit electrical signals. The transmission of a signal is the result of a neuronal *action potential* or *spike*.[1]

Definition 1 A **spike** occurs when a neuron is stimulated to a certain threshold level. At the moment a neuron spikes, an electrical pulse travels down the *axon* fibers to be received by nearby neurons. The generation of these electrical signals is an all-or-nothing event. In other words, spikes themselves do not vary in size or intensity. There is a single threshold level of stimulation that, if reached, causes a spike, otherwise no spike occurs and the signal propagates no further. The implication here is that the *spike frequency* or *rate* encodes the information being transmitted throughout the nervous system.

A typical neuron structure and the function of each component part are illustrated in Fig. 1. Nerve fibers called *dendrites* branch out from the neuron cell body or *soma*. Electrochemical signals travel down the *axon* cable and are propagated to

Fig. 1 A typical neuron structure and the function of each component part

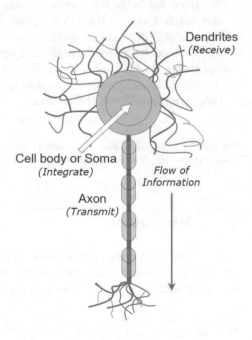

Dendrites
(Receive)

Cell body or Soma
(Integrate)

Flow of
Information

Axon
(Transmit)

---

[1] We will use the term *spike* for the remainder of the chapter, which is synonymous with the firing of an action potential.

neighboring neurons via *synapses*. A synapse is the junction where the axon of a transmitting or *presynaptic* neuron meets with the dendrite of a receiving neuron, called the *postsynaptic* neuron. Synaptic connections allow for the electrical pulses from generated spikes to travel rapidly from neuron to neuron.

1.1 Table of Symbols and Notation

Throughout the chapter, various symbols denoting model parameters will be mentioned and may be unfamiliar. Each will be defined as they are introduced, and however a table of these has been provided at the end of the chapter for your reference. See Table 1 on page 25.

1.2 Exercises and Programming Challenge Problems

In this chapter we present spiking neuron models that describe the mechanisms underlying the spiking behavior of neurons. We begin by introducing and defining the biophysical properties of a neuron. We then explore the relationships between

Table 1 Model parameters: symbols and units

General electronics and circuits	
Voltage/electrical potential	Symbol: V Units: mV (millivolt)
Charge	Symbol: Q Units: C (coulomb)
Current	Symbol: I Units: nA (nanoampere)
Resistance	Symbol: R Units: MΩ (megaohm)
Capacitance	Symbol: C Units: nF (nanofarad)
Conductance	Symbol: g Units: mS (millisiemen)
Neuroelectronics	
Membrane potential	Symbol: V Units: mV (millivolt)
Reversal potential	Symbol: E Units: mV (millivolt)
Injected current	Symbol: I_e Units: nA (nanoampere)
Total membrane current	Symbol: I_m Units: nA (nanoampere)
Membrane resistance	Symbol: R_m Units: MΩ (megaohm)
Membrane capacitance	Symbol: C_m Units: nF (nanofarad)
Specific membrane resistance	Symbol: r_m Units: M$\Omega \times$ mm^2
Specific membrane capacitance	Symbol: c_m Units: nF / mm^2
Membrane surface area	Symbol: A Units: mm^2 (square millimeter)
Membrane time constant	Symbol: τ_m Units: ms (millisecond)
Ions	
Potassium	Symbol: K^+
Sodium	Symbol: Na^+
Calcium	Symbol: Ca^{2+}

these properties and discuss how they are incorporated into spiking neuron models. This is followed by a derivation of the integrate-and-fire model, the first of several models presented in the chapter.

Throughout the chapter, there will be exercises meant to solidify your under-standing of the material up to that point. These should be considered as **required**, as some concepts may be introduced in the exercises themselves. It is highly recommended to take the time to understand the exercises before moving on. The answers to each exercise can be found at the end of the chapter.

Beginning with Sect. 3, each model we present will have a set of **Coding Challenge Problems**. While these problems are coded in Python, they do not require any sophisticated programming skills to complete. The necessary Python code has already been prepared and needs only to be executed to see the results, all of which is thoroughly explained in the content provided. Of course, you are certainly encouraged to read through the documented code in a manner consistent with any programming experience you may have.

For all coding challenge problems, visit
https://github.com/mbohling/spiking-neuron-model

2   Modeling the Neuron

The spiking behaviors of the model neurons considered in this chapter are dependent on a single variable called the *membrane potential*. Neurons modeled in this way are referred to as *single-compartment* neurons, a simplified representation of a biological neuron that allows for computing the membrane potential as a function of several biophysical properties of the neuron such as the *membrane resistance*.

Definition 2 The **membrane potential** of a neuron is the difference in voltage or electrical potential between the inside of the cell membrane and the surrounding extracellular medium. This voltage value, denoted V, is spread across the membrane surrounding the neuron cell body and is measured in millivolts (mV).

Definition 3 The **membrane resistance** of a neuron measures the level of opposi-tion to the flow of electrical current through the neuronal membrane. This quantity is denoted as R_m and is measured in megaohms (MΩ).

For the purposes of the models discussed in this chapter, the membrane resis-tance, R_m, is considered to be constant.[2] This assumption allows us to employ *Ohm's law*, a well-known relationship used to describe the behavior of simple series electrical circuits typically expressed as

---

[2] In a biological neuron, the membrane resistance varies over the surface area, and thus for simplicity, R_m is considered to be constant and represents the average resistance across the neuronal surface.

$$V = IR, \tag{1}$$

where the electrical potential of the circuit, V, is directly proportional to the flow of current through the circuit, I, and the resistance, R, is defined as being the constant of proportionality [12]. In neurons, an electrical current, I, that passes through the neuron cell membrane may be generated from various sources such as an external sensory stimulus or random *noise* originating from activity located in other regions of the brain. In an experimental setting, *injected current* is delivered directly into the neuron via an electrode. For the models in this chapter, I_e is considered to be an injected current. Therefore, Eq. (1) can be translated to describe the behavior of a neuron when expressed as

$$\Delta V = I_e R_m. \tag{2}$$

Thus, by Eq. (2) and definition 3, it follows that the membrane resistance of a neuron serves as the constant of proportionality and determines the magnitude of change in membrane potential.

Exercise 1 The *membrane surface area* of a neuron may range from 0.01 to 0.1 mm^2 and is denoted A. The membrane resistance per unit area along the neuronal surface area is the *specific membrane resistance*, denoted r_m, where $r_m = R_m A$ and is measured in M$\Omega \times$ mm^2.

(a) Consider a neuron, N_1, with a surface area of 0.05 mm^2 and a membrane resistance of 18 MΩ. Compute the specific membrane resistance r_m.
(b) Now consider a second neuron, N_2, with surface area 0.075 mm^2 and a specific membrane resistance membrane resistance $r_m = 0.9375$ M$\Omega \times$ mm^2. Compute the membrane resistance R_m.
(c) Using Eq. (2), compute ΔV for neurons N_1 and N_2 where an injected current $I_e = 2.5$ nA (or nanoamperes, the basic unit we use to measure the strength of an electrical current) is applied. Which neuron undergoes the larger change in membrane potential?

Exercise 2 The *resting potential* of a neuron, denoted V_{rest}, is the equilibrium membrane potential in the absence of any external stimulus. In spiking neuron models, V_{rest} is conventionally set to -70 mV. Consider a model neuron at rest with membrane resistance $R_m = 25$ MΩ. How much external current is required to hold the membrane potential at -50 mV?

As electrical current enters a neuron, the cell membrane acts as a capacitor and builds up *charge*, denoted Q, which is distributed across the membrane. The amount of charge stored, measured in coulombs (C), is directly proportional to the membrane potential. The ratio between the charge, Q, and the membrane potential, V, is determined by the *membrane capacitance*.

Definition 4 The **membrane capacitance** is the ratio between the charge, Q, stored across the neuronal membrane and the membrane potential, V. This quantity is denoted as C_m and measured in nanofarads (nF).

Exercise 3 The *membrane time constant*, denoted τ_m, where $\tau_m = R_m C_m$, sets the basic time scale for changes in the membrane potential and typically ranges from 10 to 100 ms (milliseconds).

(a) Consider a model neuron, N_1, with membrane resistance, $R_m = 9$ MΩ, and a membrane capacitance of $C_m = 5$ nF. Compute the membrane time constant τ_m.
(b) Consider a second model neuron, N_2, with $R_m = 9$ MΩ and $C_m = 3$ nF. If both N_1 and N_2 are at rest with $V_{rest} = -70$ mV (defined in exercise 2), and an injected current of equal strength is applied to both neurons simultaneously, which neuron will spike first? [Note: Assume the injected current is strong enough to cause both neurons to spike.]

2.1 The Neuronal Spike

Many neuron models employ a *spiking mechanism* defined by a *reset condition*. This condition describes the behavior of the neuron at the moment a spike occurs. Figure 2 depicts the evolution of the membrane potential of a neuron before, during, and after a spike occurs. The reset condition is given by the following rules:

1. If the membrane potential, V, reaches the threshold value, V_{th}, then a spike occurs, and V increases to V_{spike}. An increase in membrane potential is called *depolarization*.
2. Immediately following the spike, the membrane potential decreases until it reaches the V_{reset} value. A decrease in membrane potential is called *repolarization*.
3. The membrane potential then returns to its resting potential, V_{rest}. This process is called *hyperpolarization*. The short period of time before the neuron membrane potential returns to its resting potential value is called the *refractory period* (discussed in detail in Sect. 4).

2.2 Ionic Concentrations, Conductance, and Reversal Potentials

The electrical potentials inside and outside of a neuron are determined by concentrations (or the amount) of charged particles called *ions*.

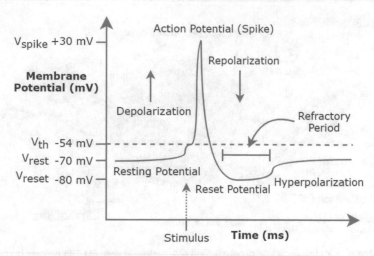

Fig. 2 A neuronal spike where membrane potential is plotted as a function of time

Definition 5 An **ion** is an atom that has an electrical charge. In this chapter, we consider two types of positively charged ions commonly found in the biological makeup of a neuron: *potassium* ions, written K^+, and *sodium* ions, written Na^+.

The difference in ionic concentrations determines whether an ion moves in or out of the neuron through channels in the membrane. The relationship between ionic concentrations and the inward and outward flows of electrical current through the neuron cell membrane can best be explained by Fig. 3.

The expression used to measure the membrane current per unit area due to channels that conduct a particular ion type, i, is given by

$$I_i = g_i(V - E_i), \tag{3}$$

where g_i is the *conductance* per unit area and E_i is the *reversal potential*.

Definition 6 The **reversal potential** of an ion-conducting channel is the membrane potential at which point there is no net flow of electrical current across the cell membrane.

Definition 7 The electrical **conductance** of an ion channel is a measure of the ease through which electric current is able to flow through the neuronal membrane. The conductance determines the strength of the current and is the reciprocal value of the membrane resistance. This quantity is measured in millisiemens (mS).

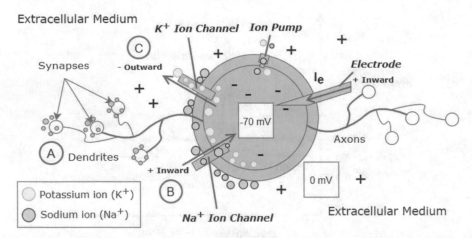

Fig. 3 Molecular view depicting depolarization of a neuron at rest. (**a**) The process begins when the dendrites receive an electrochemical signal from the presynaptic neuron. (**b**) The signal starts to increase the membrane potential by the opening of Na$^+$-conducting channels producing a positive-inward current. (**c**) If the signal is strong enough, the membrane potential reaches a threshold value of −54 mV, resulting in a spike to upwards of +30 mV. Thus, the potential inside the cell is now positive relative to the extracellular medium, and positively charged K$^+$ ions are repelled outside of the neuron to the medium with less positive electrical potential

Exercise 4 In spiking neuron models, the *total membrane current* per unit area is the sum of all ionic currents

$$I_m = \sum_i g_i(V - E_i). \tag{4}$$

Consider a model neuron with a sodium current with conductance $g_{Na} = 1.2\,\mathrm{mS/mm^2}$ and $E_{Na} = 50\,\mathrm{mV}$ and a potassium current with $g_K = 0.36\,\mathrm{mS/mm^2}$ and $E_K = -87\,\mathrm{mV}$. Use Eq. (4) to compute the total membrane current per unit area at the moment $V = 0\,\mathrm{mV}$. [Note: The summation symbol Σ_i in Eq. (4) indicates we are adding one term for each ionic current denoted by the subscript i. In this particular case, we can translate the summation as $I_m = \sum_i g_i(V - E_i) = g_{Na}(V - E_{Na}) + g_K(V - E_K)$.]

Exercise 5 Observe that Eq. (3) may be rewritten as

$$g_i = \frac{I_i}{V - Ei}. \tag{5}$$

If we assume a constant ionic current, I_i, is passing through a channel with reversal potential, E_i, then we can measure electrical conductance as a function of the membrane potential. Consider a neuron with a constant sodium current $I_{Na} = -140.4\,\mathrm{nA}$ passing through a sodium-conducting channel with $E_{Na} = 52\,\mathrm{mV}$ and a constant potassium current $I_K = 4.32\,\text{nA}$ passing through a potassium-conducting channel with $E_K = -77\,\text{mV}$.

(a) Compute the conductance g_{Na} when $V = -65\,\text{mV}$.
(b) Compute the conductance g_K when $V = 0\,\text{mV}$.
(c) If the membrane potential $V > E_K$, is the current I_K positive or negative? Are K^+ ions flowing into or out of the neuron? Explain. [Hint: K^+ ions are positively charged.]

---

3 Integrate-and-Fire Spiking Neuron Model

To derive the membrane equation for a neuron, we refer back to the following relationship between its membrane capacitance, C_m, the net amount of charge it stores, Q, and the membrane potential (see Definition 4).

$$Q = C_m V. \tag{6}$$

Just like the membrane resistance, the membrane capacitance varies over the surface area; thus for simplicity, C_m is considered to be constant and represents the average capacitance across the neuronal surface. Thus, if we differentiate equation (6) with respect to time, we obtain

$$\frac{dQ}{dt} = C_m \frac{dV}{dt}. \tag{7}$$

The rate of change of the amount of charge stored is equal to the current passing into the neuron. When accounting for external sources of current, I_e, Eq. (7) becomes

$$C_m \frac{dV}{dt} = -I + I_e, \tag{8}$$

where I is the total membrane current. By convention, the external current entering the neuron is defined as *positive-inward*, and the membrane current I_m is defined as *positive-outward*, which explains the different signs [1]. Dividing equation (8) by the neuronal surface area thus yields

$$c_m \frac{dV}{dt} = -I_m + \frac{I_e}{A}, \tag{9}$$

where I_m represents the membrane current per unit area and $c_m = C_m/A$ is the *specific membrane resistance*. If I_m is expressed in terms of a single *leakage*[3]

---

[3] Leakage channels are responsible for maintaining the resting membrane potential. This is because the reversal potential, E_L, serves as V_{rest} in many simulations.

ion channel with conductance, g_L, and reversal potential, E_L, then Eq. (9) may be written as

$$c_m \frac{dV}{dt} = -g_L(V - E_L) + \frac{I_e}{A}. \tag{10}$$

Equation (10) can be simplified to a more convenient expression by multiplying both sides by the specific membrane resistance r_m. This results in

$$r_m c_m \frac{dV}{dt} = -r_m g_L(V - E_L) + r_m \frac{I_e}{A}. \tag{11}$$

Recall the following relationships:

- $r_m = R_m A$ (see Exercise 1).
- By definition, the membrane time constant $\tau_m = r_m c_m$ (see Exercise 3).
- By Definition 7, the resistance is the reciprocal value of conductance. In this case, $r_m = 1/g_L$.

Thus, Eq. (11) can be rewritten as

$$\tau_m \frac{dV}{dt} = E_L - V + R_m I_e. \tag{12}$$

Additionally, to capture the spiking behavior described in Fig. 2, the following reset condition is imposed on V: if the membrane potential, V, reaches the threshold value, V_{th}, at time t, then $V(t) \leftarrow V_{spike}$ and $V(t + 1) \leftarrow V_{reset}$.[4]

Equation (12), along with reset condition on V_{th}, is referred to as the *integrate-and-fire* model [3]. It is sometimes referred to as the *leaky* integrate-and-fire model because of the leakage term.

The integrate-and-fire model is still considered an extremely useful description of neuronal activity despite the fact that it was originally proposed over a hundred years ago by Louis Lapicque in 1907. Recent extensions of the model include the *exponential integrate-and-fire neuron model* proposed by Fourcaud et al. This variant reproduces the dynamics of a simple conductance-based model extremely well and shows how an intrinsic neuronal property determines the speed with which neurons can track changes in input [4]. This has been built upon further by Gerstner and Brette who more recently proposed the *adaptive exponential integrate-and-fire model*. The model they present is capable of describing well-known neuronal

---

[4] The arrow notation $\leftarrow$ should be thought of as assigning a particular voltage value, in this case V_{spike} and V_{reset}, to the value V, at time t. For example, if the membrane potential reaches the threshold value, V_{th} set to -54 mV at time $t = 3$ ms, then $V(t) = V(3) = V_{spike}$ and $V(t + 1) = V(4) = V_{reset}$. The idea of *assigning* these values allows us to think about the evolution of changing voltage values of a neuron over time.

firing patterns, e.g., adapting, bursting, delayed spike initiation, initial bursting, fast
spiking, and regular spiking [5].

3.1   Numerical Implementation of the Model

To compute and plot the evolution of the membrane potential, we employ *Euler's
method* of numerical integration [2]. To simulate the integrate-and-fire model, we
start by rewriting equation (12) as

$$\frac{dV}{dt} = \frac{1}{\tau_m}(E_L - V + R_m I_e). \tag{13}$$

Now assume that the initial value of the membrane potential (at time t_0) is V_0 and
that t_1 is some time after t_0 and very close to t_0. Let $V_1 = V(t_1)$, then we can
numerically approximate equation (13) as follows:

$$\frac{V_1 - V_0}{\Delta_1} = \frac{1}{\tau_m}(E_L - V_0 + R_m I_e) \tag{14}$$

or equivalently as

$$V_1 = V_0 + \frac{\Delta_1}{\tau_m}(E_L - V_0 + R_m I_e), \tag{15}$$

where $\Delta_1 = t_1 - t_0$ is called a *time step*.

Next, we choose another time t_2 very close to t_1 to obtain V_2 (i.e., $V(t_2)$) by
using a numerical scheme similar to Eq. (15). We can continue in this fashion for
t_3 very close to t_2, t_4 very close to t_3, and so on. Thus, Euler's method numerical
approximation for Eq. (13) can be summarized as follows:

$$V_n = V_{n-1} + \frac{\Delta_n}{\tau_m}(E_L - V_{n-1} + R_m I_e), \tag{16}$$

where $\Delta_n = t_n - t_{n-1}$. The smaller the value of Δ_n is, the more accurate the result is,
and however small time steps increase simulation time as more iterations of Eq. (16)
are required.

This procedure is summarized in Algorithm 1.

Algorithm 1: INTEGRATEANDFIRE()

---

for $t \leftarrow t_0$ **to** $t_{final} - 1$

$$
\textbf{do}
\begin{cases}
dV \leftarrow \dfrac{dt}{\tau_m}(E_L - V(t) + R_m I_e) \\
V(t+1) \leftarrow V(t) + dV \\
\begin{cases}
\textbf{if } V(t+1) \geq V_{th} \\
\\
\textbf{then } \begin{cases} V(t) \leftarrow V_{spike} \\ V(t+1) \leftarrow V_{reset} \end{cases}
\end{cases}
\end{cases}
$$

Exercise 6 Let $t_0 = 0$, $t_{final} = 3$ ms, $dt = 1$ ms, $\tau_m = 10$ ms, $E_L = -70$ mV, $V(t_0) = -55$ mV, $R_m = 10$ MΩ, $I_e = 2.0$ nA, and $V_{th} = -54$ mV. Given these parameter values, the simulation time is 3 ms with a time step of 1 ms. Using Algorithm 1, will the neuron spike during the simulation? What about if $I_e = 2.1$ nA?

3.2   Coding Challenge Problems

For the following set of coding challenge problems, visit
https://github.com/mbohling/spiking-neuron-model
    Select the link next to the *Integrate-and-Fire Spiking Neuron Model*, and follow the instructions found in the corresponding Google Colab Notebook.

Challenge Problem 1

(a) Begin by carefully reading the section titled **How it Works**. Then, execute the code block *Initialize Setup*.
(b) Complete the section titled **Walkthrough** by running each code block in the section. This section is followed by the **Full Code** of the Python implementation of the model.
(c) Navigate to the last section of the notebook titled **Coding Challenge Problems**, and execute the code block *Run Simulation*. Use the controls above the plot to complete the following:

(continued)

(i)   Set the injected current, I_e, to 1.60 nA, the membrane resistance, R_m, to 10 MΩ, and membrane capacitance, C_m, to 1 nF. Does the neuron spike at all during the simulation?

(ii)   Now, set I_e to 1.61 nA. Does the neuron spike now? If so, explain why.

(iii)   Finally, explain how the spiking behavior changes as you increase the membrane capacitance. What parameter in Eq. (12) is responsible for the change in behavior?

4   Integrate-and-Fire with Spike-Rate Adaptation

While the integrate-and-fire model provides us with a simplified view of neural spiking behavior, it does so by ignoring many properties exhibited by biological neurons. An important neuronal property ignored by Eq. (12) is *spike-rate adaptation*. Spike-rate adaptation is the reduction of a neuron's ability to spike while undergoing a stimulus of constant intensity. This occurs during a *refractory period*.

Definition 8 During repetitive spiking of a neuron, the probability that the neuron spikes is significantly reduced, a property called **refractoriness**. This lasts for a short period of time, called the **refractory period** (see Fig. 2).

Refractoriness can be incorporated into a spiking neuron model in a few different ways. For example, we could use the same membrane equation (12) with an additional rule that after a spike, we switch off the external current, I_e, for a short period of time during the simulation before switching it back on. In a similar fashion, we could also increase the voltage threshold level, V_{th}, requiring a higher membrane potential in order to spike.

While those solutions are viable, a more realistic implementation of refractoriness is done by extending the membrane equation to include an additional *spike-rate adaptation conductance* term. The resulting model is expressed as follows:

$$\tau_m \frac{dV}{dt} = E_L - V - r_m g_{sra}(V - E_K) + R_m I_e, \tag{17}$$

where the additional conductance term, $r_m g_{sra}(V - E_K)$, is modeled as a K$^+$ conductance with a reversal potential E_K, typically in the range between -90 mV and -70 mV (therefore, it has an inhibitory effect on the membrane potential). Thus, when this conductance is activated, the neural membrane depolarizes. The reset condition for the leaky integrate-and-fire model still applies here. Additionally, the adaptation conductance, g_{sra}, adheres to the following rules:

1. At each spike occurrence, $g_{sra} = g_{sra} + dg_{sra}$.
2. In the absence of spikes, the amount dg_{sra} decays exponentially to 0 according to the time constant, τ_{sra}, by the following equation:

$$\tau_{sra} \frac{dg_{sra}}{dt} = -g_{sra}. \tag{18}$$

This procedure is summarized in Algorithm 2.

Algorithm 2: INTEGRATEANDFIRE_SRA()

---

for $t \leftarrow t_0$ **to** $t_{final} - 1$

do $\begin{cases} dV \leftarrow \dfrac{dt}{\tau_m}(E_L - V(t) - r_m g_{sra}(V(t) - E_K) + R_m I_e) \\ V(t+1) \leftarrow V(t) + dV \\ \textbf{if } V(t+1) \geq V_{th} \\ \quad \textbf{then } \begin{cases} V(t) \leftarrow V_{spike} \\ V(t+1) \leftarrow V_{reset} \\ g_{sra}(t+1) \leftarrow g_{sra}(t) + dg_{sra} \end{cases} \\ \quad \textbf{else } \begin{cases} g_{sra}(t+1) \leftarrow \dfrac{dt}{\tau_{sra}}(-g_{sra}(t)) \end{cases} \end{cases}$

4.1   Coding Challenge Problems

For the following set of coding challenge problems, visit
https://github.com/mbohling/spiking-neuron-model
    Select the link next to *Integrate-and-Fire with Spike-Rate Adaptation*, and follow the instructions found in the corresponding Google Colab Notebook.

Challenge Problem 2

(a) Begin by carefully reading the section titled **How it Works**. Then, execute the code block *Initialize Setup*.

(continued)

(b) Complete the section titled **Walkthrough** by running each code block in the section. This section is followed by the **Full Code** of the Python implementation of the model.

(c) Navigate to the last section of the notebook titled **Coding Challenge Problems**, and execute the code block *Run Simulation*. Use the controls above the plot to complete the following:

(i)   Set the injected current, I_e, to 2.00 nA and dg_{sra} to 0.15 nF, and leave τ_{sra} at 100 ms. How does the spike behavior differ from the leaky integrate-and-fire simulations? [Note: The membrane equation (12) without spike-rate adaptation is called *leaky*.]

(ii)   Set I_e to 2.50 nA, dg_{sra} to 0.02 nF, and τ_{sra} to 20 ms. What is the maximum value of g_{sra} over the simulation?

(iii)   Now increase τ_{sra} to 200 ms. What is the maximum value of g_{sra} over the simulation?

(iv)   How does increasing the value of dg_{sra} affect the number of spikes throughout the simulation? Explain.

5   Hodgkin–Huxley Spiking Neuron Model

The Hodgkin–Huxley spiking neuron model follows the membrane potential description expressed in Eq. (9)

$$c_m \frac{dV}{dt} = -I_m + \frac{I_e}{A}.$$

However, the membrane current equation is given by

$$I_m = I_L + I_K + I_{Na}. \tag{19}$$

Thus, the neuron cell membrane current consists of three types of ion channels including a leakage current which is expressed as

$$I_L = g_L(V - E_L). \tag{20}$$

The other channels are *voltage-dependent* channels: one permeable to K$^+$ ions and the other permeable to Na$^+$ ions. A voltage-dependent channel consists of several small gates that open and close, classified as either *activation gates* or *inactivation gates*. The open or closed state of each gate depends on its *open probability*. The open or closed states of each channel gate are mutually exclusive. Additionally, all gates must be open in order for ions to pass through the channel.

The K$^+$ ion channel consists of four activation gates (see Fig. 4). The current is expressed as

Fig. 4 Visualizing activation gates. Each gate has open probability n. Ions may only pass through with all gates open. The probability changes according to opening and closing rate functions which depend on the membrane potential detected by the voltage sensor

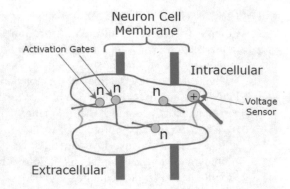

$$I_K = g_K n^4 (V - E_K),\tag{21}$$

where n represents the open probability that an activation gate is open. The Na$^+$ ion channel consists of three activation gates and one inactivation gate. The current is expressed as

$$I_{Na} = g_{Na} m^3 h (V - E_{Na}),\tag{22}$$

where m represents the open probability that an activation gate is open, and h represents the open probability that the inactivation gate is open. Thus, Eq. (19) may be rewritten as

$$I_m = g_L(V - E_L) + g_K n^4 (V - E_K) + g_{Na} m^3 h (V - E_{Na}).\tag{23}$$

Equation (23) describes the membrane current of the Hodgkin–Huxley model neuron [6]. The gating variables n, m, and h vary between 0 and 1 according to opening and closing rate functions, $\alpha(V)$ and $\beta(V)$, respectively. For each gating variable x, these are computed as

$$\tau_x(V) \frac{dx}{dt} = x_\infty(V) - x,\tag{24}$$

where

$$\tau_x(V) = \frac{1}{\alpha_x(V) + \beta_x(V)}\tag{25}$$

and

$$x_\infty(V) = \frac{\alpha_x(V)}{\alpha_x(V) + \beta_x(V)}.\tag{26}$$

This procedure is summarized in Algorithm 3.

Algorithm 3: HODGKINHUXLEY()

---

for $t \leftarrow t_0$ **to** $t_{final} - 1$

do
$$
\begin{cases}
\textbf{for } var \leftarrow [n, m, h] \\
\textbf{do }
\begin{cases}
\tau_{var} \leftarrow \dfrac{1}{\alpha_{var}(V(t)) + \beta_{var}(V(t))} \\[2ex]
var_{\infty} \leftarrow \dfrac{\alpha_{var}(V(t))}{\alpha_{var}(V(t)) + \beta_{var}(V(t))} \\[2ex]
dvar \leftarrow \dfrac{dt}{\tau_{var}}(var_{\infty} - var(t)) \\[2ex]
var(t+1) \leftarrow var(t) + dvar
\end{cases} \\[2ex]
i_m(t+1) \leftarrow \overline{g}_L(V(t) - E_L) + \overline{g}_K n(t+1)^4 (V - E_K) + \overline{g}_{Na} m(t+1)^3 \\[2ex]
h(t+1)(V - E_{Na}) \\[2ex]
dV \leftarrow dt\left(-i_m(t+1) + \dfrac{I_e}{A}\right) \\[2ex]
V(t+1) \leftarrow V(t) + dV
\end{cases}
$$

The Hodgkin–Huxley model is an extremely important spiking neuron model and the original formulation of a *conductance-based* spiking neuron model. These models include conductance terms related to our discussion of ion-conducting channels in Sect. 2.2 and are the most common formulation used in neuronal models [7]. One of the more recent extensions of the model is the *Thermodynamic Hodgkin–Huxley model*. This model describes voltage dependence empirically as well but can be construed to be more physically plausible, as they constrain and parameterize their fit with thermodynamic principles of transition state theory [8].

5.1 Coding Challenge Problems

For the following set of coding challenge problems, visit
https://github.com/mbohling/spiking-neuron-model
Select the link next to the *Hodgkin–Huxley Spiking Neuron Model*, and follow the instructions found in the corresponding Google Colab Notebook.

Challenge Problem 3

(a) Begin by carefully reading the section titled **How it Works**. Then, execute the code block *Initialize Setup*.
(b) Complete the section titled **Walkthrough** by running each code block in the section. This section is followed by the **Full Code** of the Python implementation of the model.
(c) Navigate to the last section of the notebook titled **Coding Challenge Problems**, and execute the code block *Run Simulation*. Use the controls above the plot to complete the following:

   (i)   Leave the values at their default values. The injected current $I_e = 1.75$ nA is switched on at 5 ms and switched off at 10 ms. At approximately what time t does the membrane current, i_m, first become positive? At this moment, what are the approximate values of the activation variable, n, and the membrane potential, V? Explain this behavior.
   (ii)   Set the injected current $I_e = 0.50$ nA, and set the range so that the current is switched on at 1 ms and off at 2 ms. Now, increase the injected current by 0.05 nA. Notice the drastic change in the behavior of the membrane potential. Explain how this small increase in current causes this.

6   Integrate-and-Fire with Synaptic Conductance

The final spiking neuron model discussed in this chapter is similar to the spike-rate adaptation model in that it is effectively another extension of the integrate-and-fire model. To incorporate synaptic conductance into Eq. (12), an additional current term so that the new membrane equation becomes

$$\tau_m \frac{dV}{dt} = E_L - V - r_m g_s P_s (V - E_s) + R_m I_e \tag{27}$$

where g_s is the synaptic conductance, E_s is the reversal potential due to synaptic conductance, and P_s denotes the *open channel probability*, or the probability that a postsynaptic ion channel opens given that a spike was fired by the presynaptic neuron.

A synaptic conductance is the product of synaptic transmission; that is, molecules called neurotransmitters are released across a small gap at the synapse called the *synaptic cleft*. These transmitters bind to receptors on the postsynaptic neuron. Depending on the type of neurotransmitter and corresponding receptor, the binding reaction leads to the opening ion channels (see Fig. 5).

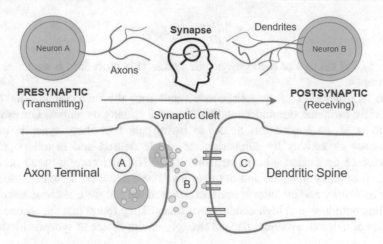

Fig. 5 The synaptic transmission mechanism. (**a**) The electrical signal travels down the axon cable of a presynaptic neuron to the terminal points at synapses between neighboring postsynaptic neurons. This causes synaptic vesicles to release neurotransmitters across the synaptic cleft. (**b**) Neurotransmitters released across the synaptic cleft bind to receptors located on the dendrites on a postsynaptic neuron. (**c**) Once a neurotransmitter binds with a receptor a signal is sent to the neuron causing ion channels to open

Definition 9 Molecular chemical messengers known as **neurotransmitters** carry an electric signal across a synapse. They are stored in containers called *synaptic vesicles* and are released across the synaptic cleft when a spike reaches the axon terminal.

When the open probability P_s is greater than 0, electrical current is able to flow into and out of the neuron. The parameter, P_s, is computed using following expression:

$$P_s(t) = \frac{P_{max}t}{\tau_s} \exp\left(1 - \frac{t}{\tau_s}\right), \tag{28}$$

where P_{max} is the maximum value of P_s and τ_s is a time constant that controls how fast the probability P_s evolves with time. The model described by Eq. (28) assumes a single presynaptic spike. We can model multiple spikes over a simulation by adding another pair of differential equations for a parameter denoted S_{Pre}. The change in open channel probability is given by

$$\tau_s \frac{dP_s}{dt} = e P_{max} S_{pre} - P_s \tag{29}$$

with $e = \exp(1)$, and

$$\tau_s \frac{dS_{pre}}{dt} = -S_{pre},$$ (30)

where S_{pre} is set to 1 at each presynaptic spike. This procedure is summarized in Algorithm 4.

The incorporation of synaptic conductance into the integrate-and-fire model has greatly furthered our understanding of neural spiking dynamics. For example, Stefano et al. have used this model to investigate how these synaptic models affect network activity by comparing the single neuron and neural population dynamics of conductance-based networks (COBNs) and current-based networks (CUBNs) of leaky integrate-and-fire neurons. These networks were endowed with sparse excitatory and inhibitory recurrent connections and were tested in conditions including both low- and high-conductance states. They found that the second-order statistics of network dynamics depend strongly on the choice of synaptic model [9].

Algorithm 4: INTEGRATEANDFIRE_SYN()

---

for $t \leftarrow t_0$ **to** $t_{final} - 1$

do $\begin{cases} \begin{cases} \textbf{if } \text{didSpike} \\ \quad \textbf{then } \begin{cases} S_{pre} \leftarrow 1 \end{cases} \\ \quad \textbf{else } \begin{cases} dS_{pre} \leftarrow \dfrac{dt}{\tau_s}(-S_{pre}) \\ S_{pre} \leftarrow S_{pre} + dS_{pre} \end{cases} \\ S_{pre}(t+1) \leftarrow S_{pre} \\ \\ dP_s \leftarrow \dfrac{dt}{\tau_s}(eP_{max}S_{pre}(t+1) - P_s(t)) \\ P_s(t+1) \leftarrow P_s(t) + dP_s \\ \\ dV \leftarrow \dfrac{dt}{\tau_m}(E_L - V(t) + r_m\overline{g}_s P_s(t+1)V(t)) \\ V(t+1) \leftarrow V(t) + dV \\ \\ \begin{cases} \textbf{if } V(t+1) \geq V_{th} \\ \quad \textbf{then } \begin{cases} V(t) \leftarrow V_{spike} \\ V(t+1) \leftarrow V_{reset} \\ \text{didSpike} \leftarrow true \end{cases} \end{cases} \end{cases}$

6.1   Coding Challenge Problems

For the following set of coding challenge problems, visit
https://github.com/mbohling/spiking-neuron-model
    Select the link next to *Integrate-and-Fire with Synaptic Conductance*, and follow
the instructions found in the corresponding Google Colab Notebook.

Challenge Problem 4

(a) Begin by carefully reading the section titled **How it Works**. Then, execute
    the code block *Initialize Setup*.
(b) Complete the section titled **Walkthrough** by running each code block
    in the section. This section is followed by the **Full Code** of the Python
    implementation of the model.
(c) Navigate to the last section of the notebook titled **Coding Challenge
    Problems**, and execute the code block *Run Simulation*. Use the controls
    above the plot to complete the following:

   (i)   Observe that we *artificially* trigger presynaptic spikes at t =0, 50,
   150, 190, 300, 320, 400, and 410 ms. Explain what you see in terms of
   these spike times. [Hint: Look at the time interval between presynaptic
   spikes over the simulation.]
   (ii)   Explain what happens during the simulation as you increase the
   membrane time constant τ_m.

---

7   Conclusion

The material presented in this chapter provides a solid foundation of spiking neuron
models and a thorough introduction to understanding the basis of communication
between the central nervous system and the rest of the body. We began our
discussion with a short presentation defining the major physical components
and biophysical properties of a typical neuron. We explored the relationships
between these properties and discussed in detail how they are incorporated into
spiking neuron models. The remaining sections were devoted to the mathematical
derivations of several introductory models, including the integrate-and-fire model,
spike-rate adaptation, synaptic conductance, and conductance-based models such as
the Hodgkin–Huxley model.
    At the end of each model presented, we illustrated their relevance and importance
by mentioning both theoretical and mathematical extensions that have since been
formulated to account for further complexities of neuronal behavior. In the next
section, we present a few examples of some open research areas building on the

chapter material. We also introduce more advanced concepts such as the facilitation and depression of synapses and other conductance-based models allowing for more accurate and realistic simulations of neuronal spiking behavior.

8 Suggested Research Projects

Research Project 1 Investigate short-term synaptic facilitation and depression in terms of release probability, P_{rel}, during synaptic transmission.

Synaptic plasticity is a biological phenomenon where the *synaptic strength* changes over time. The strength of a synapse can be modeled in various ways, including the probability of transmitter release, denoted P_r [10]. In Sect. 6, we discussed synaptic conductance where the membrane equation was given by Eq. (27). This model makes the implicit assumption that during synaptic transmission, the probability that neurotransmitters will be released at the moment of a presynaptic spike is $P_r = 1$. Adjust equation (27) as follows:

$$\tau_m \frac{dV}{dt} = E_L - V - r_m g_s P(V - E_s) + R_m I_e, \tag{31}$$

where $P = P_s P_r$. We model P_r as the average steady-state release probability. Synapses may exhibit facilitation or depression depending on how we modify the probability of transmitter release P_r at presynaptic spike times. We can model a facilitating synapse by modifying P_r at each presynaptic spike as

$$P_r = P_r + F(1 - P_r),$$

where the parameter $F \in [0, 1]$ represents the degree of facilitation. Similarly, we can model a depressing synapse by modifying P_r at each presynaptic spike as

$$P_r \leftarrow P_r D$$

where the parameter $D \in [0, 1]$ represents the degree of depression.

(a) Assume that presynaptic spike times are modeled as a Poisson spike train with firing rate r. Use the following expression to model P_r such that, on average, P_r decays to P_0 between presynaptic spikes

$$P_r = P_0 + (P_r + F(1 - P_r) - P_0)\frac{r\tau_P}{1 + r\tau_P}, \tag{32}$$

(continued)

where solving for P_r gives

$$P_r = \frac{P_0 + rF\tau_P}{1 + rF\tau_P}. \tag{33}$$

Using various values of P_0, F, and τ_P on the order of 10 ms, plot the values of the release probability, P_r, and the rate of synaptic transmission, rP_r, as a function of the firing rate of a Poisson presynaptic spike train, r, as it varies from 0 to 100 (Hz).

(b) Run the same experiment with a synapse that exhibits depression. This can be done by replacing equation (32) by

$$P_{rel} = P_0 + (P_r D - P_0)\frac{r\tau_P}{1 + r\tau_P}, \tag{34}$$

where solving for P_r gives

$$P_r = \frac{P_0}{1 + r(1 - D)\tau_P}. \tag{35}$$

Using various values of P_0, D, and τ_P on the order of 100 ms, plot the values of the release probability, P_r, and the rate of synaptic transmission, rP_r, as a function of the firing rate of a Poisson presynaptic spike train, r, as it varies from 0 to 100 (Hz).

(c) Rerun simulations of an integrate-and-fire model neuron with synaptic conductance (as in Sect. 6), and however, this time, use the membrane equation given by Eq. (31). Explore how a Poisson presynaptic spike train with various firing rates, r, affects spiking behavior.

Research Project 2 Model thalamic relay neuron using the Connor–Stevens model with an additional Ca^{2+} current.

The *Connor–Stevens spiking neuron model* is an alternative conductance-based model of neural spike generation similar to the Hodgkin–Huxley model setup from Sect. 5. The membrane equation for the model is as follows:

$$V(t + dt) = V_\infty + (V(t) - V_\infty)\exp\left(\frac{-dt}{\tau_V}\right), \tag{36}$$

where

(continued)

$$V_\infty = \frac{\sum_i g_i E_i + \dfrac{I_e}{A}}{\sum_i g_i}$$

and

$$\tau_V = \frac{c_m}{\sum_i g_i}.$$

The membrane current equation is given by

$$I_m = g_L(V - E_L) + g_{Na}m^3h(V - E_{Na}) + g_K n^4(V - E_K) + g_A a^3 b(V - E_A) \tag{37}$$

with conductance values $g_L = 0.03\,\text{nS/mm}^2$, $g_{Na} = 12\,\text{nS/mm}^2$, $g_K = 2\,\text{nS/mm}^2$, and $g_A = 4.77\,\text{nS/mm}^2$. The reversal potentials are set as $E_L = -17\,\text{mV}$, $E_{Na} = 55\,\text{mV}$, $E_K = -72\,\text{mV}$, and $E_A = -75\,\text{mV}$ [11].

Use the initial values for each gating variable m, h, n, a, and b between 0 and 1. The corresponding opening and closing rate functions are given by

$$\alpha_m(V) = \frac{0.38(V + 29.7)}{1 - \exp(-0.1(V + 29.7))} \qquad \beta_m(V) = 15.2\exp(-0.0556(V+54.7))$$

$$\alpha_h(V) = 0.266\exp(-0.05(V + 48)) \qquad \beta_h(V) = \frac{3.8}{1 + \exp(-0.1(V + 18))}$$

$$\alpha_n(V) = \frac{0.02(V + 45.7)}{1 - \exp(-0.1(V + 45.7))} \qquad \beta_n(V) = 0.25\exp(-0.0125(V+55.7)).$$

The gating variables associated with the A-current, a and b, are given by

$$a_\infty = \left(\frac{0.0761\exp(0.0314(V + 94.22))}{1 + \exp(0.0346(V + 1.17))} \right)^{1/3}$$

$$\tau_a = 0.3632 + \left(\frac{1.158}{1 + \exp(0.0497(V + 55.96))} \right)$$

$$b_\infty = (1 + \exp(0.0688(V + 53.3)))^{-4}$$

$$\tau_b = 1.24 + \left(\frac{2.678}{1 + \exp(0.0624(V + 50))} \right).$$

(continued)

To model a thalamic relay neuron, we add an additional Ca^{2+} current given by

$$i_{Ca} = g_{Ca} M^2 H (V - E_{Ca}).$$

where $g_{Ca} = 0.13$ nS/mm^2 and $E_{Ca} = 120$ mV. Use initial values gating variables M and H between 0 and 1. Use the following to compute the evolution of these gating variables:

$$M_\infty = (1 + \exp(-(V + 57)/6.2))^{-1}$$

$$\tau_M = 0.612 + (\exp(-(V + 132)/16.7) + \exp((V + 16.8)/18.2))^{-1}$$

$$H_\infty = (1 + \exp((V + 81)/4))^{-1}$$

$$\tau_H = \begin{cases} \exp((V + 467)/66.6), & \text{if } V < -80 \text{ mV} \\ 28 + \exp(-(V + 22)/10.5), & \text{if } V \geq -80 \text{ mV}. \end{cases}$$

The change in each gating variable may be computed using Eqs. (24)–(26).

Using the values and equations above, run 500–1000 ms simulations with various values of injected current I_e. In particular, examine the spike behavior of the thalamic relay neuron using the Connor–Stevens model when $I_e = 0$, $I_e > 0$, and $I_e < 0$. For nonzero values injected current (both negative and positive), analyze how the spiking behavior changes as a function of the magnitude of this current.

Exercises: Answers

1. (a) 0.9; (b) 12.5; (c) N2
2. 0.8
3. (a) 45; (b) N1
4. −28.68
5. (a) 1.2; (b) 0.056; (c) positive, into
6. No. Yes

Acknowledgments This work is based upon work supported, in part, by the U.S. Army Research Office and the DEVCOM U.S. Army Research Laboratory under grant #W911NF-21-1-0192. This work was partially supported by the Army Research Office (Grant W911NF2110192).

References

1. Dayan, P., Abbott, L.F. (2005). *Theoretical Neuroscience: Computational and Mathematical Modeling of Neural Systems*. The MIT Press
2. Strogatz, S. H. (1994). *Nonlinear Dynamics and Chaos*. Persues Books Publishing, L.L.C.
3. Lapicque, L. (1907). *Quantitative investigations of electrical nerve excitation treated as polarization*. Biol Cybern Vol. 97(5-6):341-9
4. Fourcaud-Trocmé, N., Hansel, D., van Vreeswijk, C., Brunel, N. (2003). *How Spike Generation Mechanisms Determine the Neuronal Response to Fluctuating Inputs*. The Journal of Neuroscience, Vol. 23(37)11628-40
5. Wulfram, G., Brette, R. (2009). *Adaptive exponential integrate-and-fire model*. Scholarpedia, Vol. 4(6):8427
6. Hodgkin, A. L., and A. F. Huxley. (1952). *A Quantitative Description of Membrane Current and Its Application to Conduction and Excitation in Nerve*. The Journal of Physiology. Vol 117(4)500-544
7. Skinner, F. K. (2006). *Conductance-based models*. Scholarpedia, Vol 1(11):1408
8. Forrest, M. D. (2014). *Can the thermodynamic Hodgkin-Huxley Model of voltage dependent conductance extrapolate for temperature?* Computation, Vol. 2:47-60
9. Stefano, C., Stefano, P., Mazzoni, A. (2014). *Comparison of the dynamics of neural interactions between current-based and conductance-based integrate-and-fire recurrent networks*. Frontiers in Neural Circuits, Vol. 8
10. Oleskevich, S., Clements, J., Walmsley, B. (2000). *Release probability modulates short-term plasticity at a rat giant terminal*. The Journal of Physiology, Vol. 524(2)513-23
11. Connor, J. A., C. F. Stevens. (1971). *Prediction of Repetitive Firing Behaviour from Voltage Clamp Data on an Isolated Neurone Soma*. The Journal of Physiology, Vol. 213(1)31-53
12. O'Sullivan, Colm. (1980). *Ohm's law and the definition of resistance*. Physics Education, Vol. 15(1)237

Counting Lattice Walks in the Plane

Steven Klee

Abstract

A classical problem in enumerative combinatorics asks for the number of ways to walk from the origin $(0, 0)$ to a point (a, b) if one is allowed to take unit steps north or east. What if we were allowed to use different steps? In how many ways could we reach a given point (a, b)? Our goal in this chapter is to introduce this and related questions, accompanied by a scaffolding of motivating examples that will help the reader gain familiarity with methods of combinatorial proof. We include a review of known results and use those results to motivate a number of potential research projects.

Suggested Prerequisites *An understanding of vector addition is most important. Experience counting with binomial coefficients will be useful, but can be developed as part of the project. Familiarity with infinite series and power series will be helpful in some places but are not necessary.*

1   Introduction

Imagine you are standing at the origin of the two-dimensional xy-plane. At any moment in time, you are allowed to step one unit north, one unit east, or one unit northeast. For example, in one step, you can reach the point $(0, 1)$ by going north, the point $(1, 0)$ by going east, or the point $(1, 1)$ by going northeast. There is no

S. Klee (✉)
Seattle University Department of Mathematics, Seattle, WA, USA
e-mail: klees@seattleu.edu

© The Author(s), under exclusive license to Springer Nature Switzerland AG 2022   153
E. E. Goldwyn et al. (eds.), *Mathematics Research for the Beginning Student,*
Volume 2, Foundations for Undergraduate Research in Mathematics,
https://doi.org/10.1007/978-3-031-08564-2_6

Table 1 All possible walks from $(0, 0)$ to $(3, 2)$ using north, east, and northeast steps

NNEEE	ENENE	DEEN	ENDE	DDE
NENEE	ENEEN	EDEN	ENED	DED
NEENE	EENNE	EEDN	DNEE	EDD
NEEEN	EENEN	EEND	NDEE	
ENNEE	EEENN	DENE	NEDE	
		EDNE	NEED	

need to be efficient in your walk. You could also reach the point $(1, 1)$ by taking one step north, followed by one step east, or one step east, followed by one step north.

In how many different ways can you walk from $(0, 0)$ to some point (a, b) through a sequence of such steps? As an example, Table 1 shows all the possible walks from $(0, 0)$ to $(3, 2)$. Here, we use N to denote a north step, E to denote an east step, and D (for "diagonal") to denote a northeast step. We will call a sequence of N, E, and D steps an N/E/D *lattice walk*.

Looking at this table, we might make a few observations. There are 25 total ways to walk from $(0, 0)$ to $(3, 2)$, but the walks may have different lengths. We can break these 25 sequences into groups—there are 10 sequences of length 5, which use no diagonal steps, 12 sequences of length 4, which use one diagonal step, and 3 sequences of length 3, which use two diagonal steps. What is the significance of these numbers? What if we wanted to count the walks terminating at an arbitrary point (a, b)? Could we find a formula to count the number of such walks?

Questions of these types will be the motivation for the research project we outline in this chapter. We will use this running example of exploring N/E/D lattice walks to develop some notation and introduce the art of combinatorial proof before we delve into the formal problem statement.

The rest of this chapter is structured as follows. Section 1 provides a brief introduction to binomial coefficients and combinatorial proof techniques in the context of counting N/E/D lattice walks. It also includes some guided exercises that will help you discover some known results about lattice walks while also developing better intuition for combinatorics. In Sect. 2.1, we will formally state the problem and introduce the terminology and notation that will be relevant for the rest of the chapter. In Sect. 4, we give a survey of known results, which are the result of undergraduate research projects from the summer of 2017 [5] and 2019 [7].

In Sects. 1 and 4, we include a number of examples with data. Being able to build examples and produce data of your own is an essential step in this line of research, so Sect. 5 discusses the algorithms we developed to generate data for us and includes some coding exercises so that you can try to reproduce our data. In Sects. 6 and 6.2, we describe a number of open problems that might serve as motivation for a research project of your own. In Sect. 6, we also include a discussion of generating functions, which are a useful tool that can be used to solve combinatorial problems. This discussion is best motivated if you are already comfortable with the idea of a power series from calculus. If you have not encountered those objects yet, this section can safely be skipped.

2   Combinatorial Proof Techniques

Our goal in this section is to introduce some basic concepts of combinatorial proof in the context of the main problems we aim to study. For additional background, Tucker's *Applied Combinatorics* [16] provides a very readable introduction (Chapters 5–7 are most relevant to this project). Benjamin and Quinn's *Proofs that Really Count* [1] provides a beautiful introduction to the art of combinatorial proof.

Let us start by examining the number of N/E/D lattice walks from $(0, 0)$ to (a, b) that do not use any diagonal steps. Such a walk must use a east steps and b north steps, so we want to count the number of ways to arrange a letter E's and b letter N's in an ordered list. We use *binomial coefficients* to count the number of such arrangements.

Definition 1 The binomial coefficient $\binom{n}{k}$ counts the number of ways to choose k objects from a set of n distinct objects.

In our setting, a walk can be encoded as a list of $a+b$ letters. Once we choose the a positions where we will place E steps, the remaining b positions must correspond to N steps. So there are $\binom{a+b}{a}$ such walks. For example, when $(a, b) = (3, 2)$, we have to take 3 E steps and 2 N steps. If we choose the second, third, and fifth steps to be east steps, the resulting walk would be NEENE. There are $\binom{5}{3}$ ways to choose the positions of the three east steps.

For combinatorialists, the answer that there are $\binom{a+b}{a}$ walks is a perfectly satisfactory answer. But this answer may not be sufficient as we are exploring these ideas for the first time. What if we wanted an actual numerical answer instead? If your calculator could compute $\binom{5}{3}$, what number would it display? We can compute the value of a binomial coefficient using the formula

$$\binom{a+b}{a} = \frac{(a+b)!}{a!b!}. \tag{1}$$

Here, $n!$ ("n factorial") is the product of the numbers from 1 up to n. For example, $5! = 1 \cdot 2 \cdot 3 \cdot 4 \cdot 5$. In combinatorics, $n!$ counts the number of ways to arrange n distinct objects (like the numbers from 1 to n) in a list. The reason for this is that we have n options for the object that will go in the first spot of the list. Having done that, there are $n - 1$ remaining objects that can go in the second spot. Having done that, there are $n - 2$ remaining objects that can go in the third spot, and so on. We multiply these numbers to count the total number of ways to make all of these choices together.

But for our walks of N's and E's, the objects (letters) are not distinct. So let us pretend for a moment that they are. Imagine we have used a different color to write each east step $E_1, E_2, \ldots, E_a$—maybe E_1 is red, E_2 is blue, and E_3 is green. And similarly, we use a different color for each of our north steps as $N_1, N_2, \ldots, N_b$— maybe N_1 is orange and N_2 is purple. Now, there are $(a + b)!$ ways to rearrange the (distinct!) symbols $E_1, \ldots, E_a, N_1, \ldots, N_b$. But if we used a black and white printer that would not show us all the different colors (or subscripts), many of these rearrangements would look the same when we printed our data. For example, when $a = 3$ and $b = 2$, all of the following rearrangements would all appear as NEENE on the printed page:

$$
\begin{array}{cccc}
N_1E_1E_2N_2E_3 & N_1E_1E_3N_2E_2 & N_1E_2E_1N_2E_3 & N_1E_2E_3N_2E_1 \\
N_1E_3E_1N_2E_2 & N_1E_3E_2N_2E_1 & N_1E_1E_2N_2E_3 & N_2E_1E_3N_1E_2 \\
N_2E_2E_1N_1E_3 & N_2E_2E_3N_1E_1 & N_2E_3E_1N_1E_2 & N_2E_3E_2N_1E_1
\end{array}
$$

Here, there are $12 = 2! \cdot 3!$ rearrangements that represent the same walk. In general, for a given rearrangement, we can shuffle the E steps among themselves in $a!$ ways, and we can shuffle the N steps among themselves in $b!$ ways to get an equivalent representation of the same walk. This tells us that the $(a + b)!$ rearrangements of the symbols $E_1, \ldots, E_a, N_1, \ldots, N_b$ can be broken up into groups of size $a! \cdot b!$, with each group representing a unique walk from $(0, 0)$ to (a, b). This justifies Eq. (1).

For example, when $a = 3$ and $b = 2$, we have

$$
\binom{a + b}{a} = \binom{5}{3} = \frac{5!}{3!2!} = \frac{120}{12} = 10.
$$

This agrees with the 10 walks in Table 1 that use only north and east steps.

There is a second independent way of computing binomial coefficients that can be equally useful, especially as you are becoming friends with the binomial coefficients for the first time. The binomial coefficients can also be computed *recursively*, meaning a given binomial coefficient can be computed in terms of other binomial coefficients.

Theorem 1 *If a and b are nonnegative integers, then*

$$
\binom{a + b}{a} = \binom{a + b - 1}{a - 1} + \binom{a + b - 1}{a}. \tag{2}
$$

Proof We have already seen that the binomial coefficient $\binom{a+b}{a}$ counts the number of ways to walk from $(0, 0)$ to (a, b) using only north and east steps. We will show that the right side of Eq. (2) is a different way of counting the same set of objects.

Any walk from $(0, 0)$ to (a, b) ends with either a north step or an east step. If the last step is a north step, then the remaining steps must form a walk from $(0, 0)$ to $(a, b - 1)$. There are $\binom{a+b-1}{a}$ such walks. Similarly, if the last step is an east step, the remaining steps must form a walk from $(0, 0)$ to $(a - 1, b)$. There are $\binom{a-1+b}{a-1}$ such walks.

This means the total number of walks from $(0, 0)$ to (a, b) that use only north and east steps is equal to $\binom{a+b-1}{a-1} + \binom{a+b-1}{a}$. □

It is also worth noting that the formula in Eq. (2) can be proved algebraically with the help of Eq. (1):

$$
\begin{aligned}
\binom{a+b-1}{a-1} + \binom{a+b-1}{a} &= \frac{(a+b-1)!}{(a-1)!b!} + \frac{(a+b-1)!}{a!(b-1)!} \\
&= \frac{a \cdot (a+b-1)!}{a!b!} + \frac{b \cdot (a+b-1)!}{a!b!} \\
&= \frac{(a+b) \cdot (a+b-1)!}{a!b!} \\
&= \frac{(a+b)!}{a!b!} = \binom{a+b}{a}.
\end{aligned}
$$

We prefer the proof in Theorem 1 to this algebraic manipulation because it shows that both sides of Eq. (2) can be interpreted in terms of walking around a city. This gives us a better understanding of the combinatorial objects and makes the proof more meaningful.

The formula given in Eq. (2) is called a *recursive formula* because it allows us to compute new binomial coefficients in terms of known ones. This is useful if we need to quickly evaluate a binomial coefficient because it does not require us to perform a lot of multiplication and cancellation in a formula such as Eq. (1). Imagine that we will label each integer point (a, b) in the plane with the binomial coefficient $\binom{a+b}{a}$. We know $\binom{n}{0} = 1$ because the only way to walk to the point $(0, n)$ is to take all north steps. Similarly, $\binom{n}{n} = 1$ because the only way to walk to the point $(n, 0)$ is to take all east steps. This allows us to fill in some *initial conditions* on the left side of Fig. 1. Having done that, we can compute the number of walks terminating at (a, b) by adding the number of walks terminating at $(a-1, b)$ (the cell to the left of (a, b)) to the number of walks terminating at $(a, b-1)$ (the cell below (a, b)). Some partial information is filled in the right side of Fig. 1.

The right side of Fig. 1 shows the binomial coefficients arranged in a triangular array called *Pascal's triangle*, the *combinatorial triangle*, the *Khayyam triangle*, or *Yang Hui's triangle*. It is an object of fundamental importance in enumerative combinatorics. Before we explore more advanced counting techniques, let us pause to do some practice problems.

Exercise 1 Extend the data in Fig. 1 to determine the value of $\binom{10}{3}$. Check your answer using Eq. (1). It is worth writing out all the values of the binomial coefficients $\binom{n}{k}$ for $n \leq 10$ and becoming familiar with them.

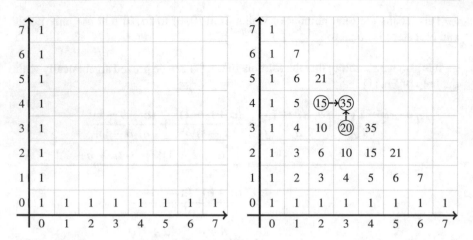

Fig. 1 Initial conditions for counting walks from $(0, 0)$ to (a, b) (left), with partially filled recursive data (right)

Exercise 2 Prove each of the following identities by interpreting binomial coefficients as the number of ways to walk around the plane using north and east steps:

1. $\binom{n}{0} + \binom{n}{1} + \cdots + \binom{n}{n} = 2^n$.
2. $\binom{b}{0} + \binom{b + 1}{1} + \binom{b + 2}{2} + \cdots + \binom{b + k}{k} = \binom{b + k + 1}{k}$.
3. $\sum_{k=0}^{r} \binom{m}{k}\binom{n}{r - k} = \binom{m + n}{r}$.

Challenge Problem 1 (If you have some coding experience.[1]) Write code in your favorite programming language or computer algebra system that will generate an array similar to the one on the right side of Fig. 1.

Now, let us move on to explore our motivating problem when some diagonal steps are allowed. Again, we will start with a concrete example. Imagine that we want to walk from $(0, 0)$ to $(5, 6)$ using exactly two diagonal steps. The two diagonal steps will ultimately contribute 2 units to the x-coordinate and 2 units to the y-coordinate, so we must also take $5 - 2 = 3$ east steps and $6 - 2 = 4$ north steps. This means we will take a total of $2 + 3 + 4 = 9$ steps, and of those steps, 2 must be diagonal, 3 must be east, and 4 must be north.

---

[1] We will discuss coding at greater length in Sect. 5.

We can think of this problem as asking for the number of ways to arrange 2 D's, 3 E's, and 4 N's in an ordered list. As we will see in Theorem 2, the number of ways to do this is

$$\frac{9!}{2!3!4!} = \binom{9}{2} \cdot \binom{7}{3} \cdot \binom{4}{4}.$$

Definition 2 Imagine you have a set of k distinct letters. The number of ways to make an ordered list with n_1 copies of the first letter, n_2 copies of the second letter, and so on, with n_k copies of the kth letter is counted by the *multinomial coefficient*

$$\binom{n_1 + \cdots + n_k}{n_1, n_2, \cdots, n_k}.$$

Theorem 2 *The multinomial coefficient $\binom{n_1+\cdots+n_k}{n_1,n_2,\cdots,n_k}$ can be computed by*

$$\binom{n_1 + \cdots + n_k}{n_1, n_2, \cdots, n_k} = \frac{(n_1 + \cdots + n_k)!}{n_1! n_2! \cdot \ldots \cdot n_k!} \tag{3}$$

$$= \binom{n_1 + \cdots + n_k}{n_1} \cdot \binom{n_2 + \cdots + n_k}{n_2} \cdot \ldots \cdot \binom{n_{k-1} + n_k}{n_{k-1}}. \tag{4}$$

Exercise 3

1. Modify the argument we gave in deriving Eq. (1) to justify Eq. (3).
2. Use the definition of the binomial coefficient (Definition 1) to justify Eq. (4). (Hint: the list has length $n_1 + \cdots + n_k$ and you need to start by choosing n_1 positions where the first letter will be placed.)
3. Use the algebraic formula for the binomial coefficients given in Eq. (1) to show that the right-hand sides of Eqs. (3) and (4) are equal.

Exercise 4

1. Show that if you are allowed to use north, east, and diagonal steps, then the number of ways to walk from $(0, 0)$ to (a, b) using exactly k diagonal steps is

$$\frac{(a + b - k)!}{(a - k)!(b - k)!k!}.$$

2. Plug in appropriate values for a, b, and k to make sure you understand why the 25 walks shown in Table 1 can be broken into groups with 10 walks of length 5, 12 walks of length 4, and 3 walks of length 3.

2.1 Walks of Minimal Length

Now, let us turn our attention to the shortest possible N/E/D walks, which we will call *minimal walks*. The left side of Fig. 2 shows the length of the shortest walk from $(0, 0)$ to (a, b) for small values of a and b when we are allowed to use north, east, and diagonal steps. The right side of Fig. 2 shows the number of walks of minimal length from $(0, 0)$ to (a, b). When we label the plane with the number of shortest walks, we see two copies of Pascal's triangle glued together along the line $y = x$. That seems pretty interesting!

Exercise 5

1. Write out all six shortest walks that terminate at the point $(2, 4)$, which is the point circled on the right side of Fig. 2.
2. Spend some time looking at Fig. 2. Make four observations about the data in either of the graphs. Can you prove your observations are correct?
3. Try to find a formula for the length of the shortest walk and the number of such walks from $(0, 0)$ to (a, b) when you are allowed to use north, east, and diagonal steps. If you need a hint or want to check your answer, see [5, Theorem 1].

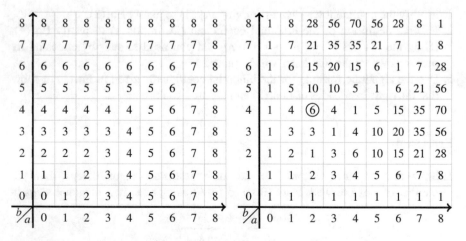

Fig. 2 The length of the shortest N/E/D walk terminating at point (a, b) (left). The number of shortest N/E/D walks terminating at point (a, b) (right)

3 The Problem

Now that we have some experience in working with N/E/D lattice walks, we are in a position to formally state the main problem we wish to investigate.

We will use $\mathbb{N}^2$ to denote the set of all vectors in the xy-plane with nonnegative integer coordinates. Let $S \subseteq \mathbb{N}^2$ be a finite set of vectors whose entries are nonnegative integers. In our motivating example of N/E/D lattice walks, $S = \{(0, 1), (1, 0), (1, 1)\}$. We view S as a set of *allowable steps*. An *S-walk* is an ordered sequence of *steps*, $\mathbf{s} = s_1, s_2, \ldots, s_L$, with $s_i \in S$ for all i. We visualize $\mathbf{s}$ as a walk in the plane, beginning at the origin and terminating at the point $s_1 + s_2 + \cdots + s_L$. We say the number of steps in the walk (L) is its *length*.

Figure 3 illustrates two walks. On the left, we see the walk $\mathbf{s} = (0, 1), (1, 0), (0, 1), (1, 0), (1, 0)$, which is encoded as NENEE in Table 1. On the right, we see the walk $\mathbf{s}' = (1, 0), (1, 1), (1, 0), (0, 1)$, which is encoded as EDEN. The walk $\mathbf{s}$ has length 5, while the walk $\mathbf{s}'$ has length 4.

The main problem that has motivated much of the existing research on this problem can be stated rather simply.

Main Question 1 Let $S \subseteq \mathbb{N}^2$ be a finite set of vectors with nonnegative integer entries.

1. In how many ways can one walk from $(0, 0)$ to (a, b) using steps from S? In other words, how many S-walks are there from $(0, 0)$ to (a, b)?
2. What is the length of the shortest S-walk from $(0, 0)$ to (a, b)? How about the longest S-walk?
3. Fix a length k. How many S-walks from $(0, 0)$ to (a, b) have length k?

We use $d(a, b; S)$ to denote the *distance* from $(0, 0)$ to (a, b) using steps in S— that is, the length of the shortest S-walk from $(0, 0)$ to (a, b). If it is impossible

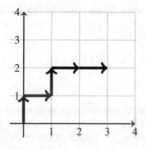
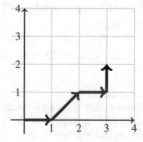

Fig. 3 Two S-walks for $S = \{(0, 1), (1, 0), (1, 1)\}$

to walk from $(0, 0)$ to (a, b) using steps in S, we say $d(a, b; S) = \infty$. We use $\mathscr{W}(a, b; S)$ to denote the set of all S-walks of minimal length from $(0, 0)$ to (a, b).

In many of our examples, the set S will contain the steps $(1, 0)$ and $(0, 1)$, which ensures that there is always an S-walk from $(0, 0)$ to any point (a, b). We say $(1, 0)$ and $(0, 1)$ are *short steps* and any remaining steps in S are called *long steps*. We use $\mathscr{W}(a, b; S, \ell)$ to denote the set of all S-walks terminating at the point (a, b) that use ℓ long steps.[2] For example, when $S = \{(1, 0), (0, 1), (1, 1)\}$, you should have seen in Exercise 5 that $|\mathscr{W}(3, 2; S, 1)| = 12$, that $|\mathscr{W}(3, 2; S, 2)| = 3$, and that $d(a, b; S) = \max(a, b)$ for any $a, b \in \mathbb{N}$.

4 Summary of Known Results

The results discussed here are the result of two undergraduate research projects from the summers of 2017 and 2019. In 2017, I worked with students Jackson Evoniuk and Van Magnan [5], who focused on walks of minimal length. In 2019, I worked with Nestor Iwanojko, Bryn Lasher, and Elena Volpi [7], who focused on walks of arbitrary length. Both projects resulted in published papers, which are freely available on the website[3] of the *Journal of Integer Sequences*. We will begin with an overview of their results.

Both papers started by exploring $S = \{(1, 0), (0, 1), (1, 1)\}$, which served as our motivating example at the beginning of this chapter. Evoniuk et al. [5, Theorem 1] studied S-walks of minimal length and established the content of Exercise 5. Iwanojko et al. [7, Theorem 1] extended their results to count S-walks of arbitrary length, giving the content of Exercise 4.

4.1 Enumeration When S Is Small

It is natural to start by considering cases where the size of S is small. Evoniuk et al. [5] gave an initial result when $S = \{(1, 0), (0, 1), (u, v)\}$, with $u, v \geq 1$.

Theorem 3 ([5, Theorem 4]) *Let* $S = \{(1, 0), (0, 1), (u, v)\}$ *with* $u, v \geq 1$, *and let* $(a, b) \in \mathbb{N}^2$. *Define* $m = \min\left(\lfloor \frac{a}{u} \rfloor, \lfloor \frac{b}{v} \rfloor\right)$.

A minimal S-walk to the point (a, b) uses exactly m steps in the (u, v)-direction. Consequently,

$$d(a, b; S) = m + a - m \cdot u + b - m \cdot v,$$

---

[2] At first, this notation may seem a bit unnatural, and it may seem more reasonable to count the number of walks of a given length instead. In our N/E/D example, a path using ℓ long steps and terminating at (a, b) has length $a + b - \ell$, so knowing the length of the walk is equivalent to knowing how many long steps it uses. It is simply more concise to fix the number of long steps. This continues to be the case in many of the sets we have explored.

[3] https://cs.uwaterloo.ca/journals/JIS/.

and

$$|\mathcal{W}(a, b; S)| = \binom{m + a - m \cdot u + b - m \cdot v}{m, a - m \cdot u, b - m \cdot v}.$$

More generally, Evoniuk et al. [5, Problem 10] explored minimal walks for sets of the form $S = \{(1, 0), (0, 1), (u, v), (v, u)\}$ with $u, v \geq 1$ arbitrary. They found that even the problem of determining the length of the shortest walk was quite difficult in this case (see [5, Figure 5]), but they did have success for the specific case[4] that $S = \{(1, 0), (0, 1), (1, 2), (2, 1)\}$ [5, Theorem 6, Theorem 9].

Iwanojko et al. [7] extended these results to enumerate all walks for sets of the form $S = \{(1, 0), (0, 1), (u, u + 1), (u + 1, u)\}$ for any $u \geq 1$. Recall that $(1, 0)$ and $(0, 1)$ are called short steps and $(u, u + 1)$ and $(u + 1, u)$ are long steps. We begin with a motivating example in the case that $u = 2$. Figure 4 shows the values of $d(a, b; S)$ for points with $0 \leq a, b \leq 11$ when $S = \{(1, 0), (0, 1), (2, 3), (3, 2)\}$. The dashed lines are spanned by the vectors $(2, 3)$ and $(3, 2)$.

Consider the values of $d(a, b; S)$ as (a, b) ranges along the diagonal where $a + b$ is fixed. For example, when $a + b = 11$, those distances are

$$(11, 11, 7, 7, 3, 3, 3, 3, 7, 7, 11, 11).$$

Here, we observe three properties of these distances. First, $d(a, b; S) = d(b, a; S)$ because the vectors in S are symmetric about the line $y = x$. Thus, we need to only consider distances $d(a, b; S)$ when $a \leq b$, or equivalently, when $a \leq \frac{a+b}{2}$. Second, aside from the run of consecutive threes at the center of this list, all other values (in this case, 11 and 7) appear in blocks of size $u = 2$. Third, the distinct values in this list (11, 7, and 3) form an arithmetic progression whose common difference is $2u = 4$ and whose initial value is $a + b = 11$. Corollary 1 below shows this pattern continues for all $u \geq 1$. We begin with a more general result.

Theorem 4 ([7, Theorem 3]) *Let $u \geq 1$ and $S = \{(1, 0), (0, 1), (u, u + 1), (u + 1, u)\}$. For any $(a, b) \in \mathbb{N}^2$, the number of S-walks terminating at (a, b) that use ℓ long steps is*

$$|\mathcal{W}(a, b; S, \ell)| = \binom{a + b - 2u\ell}{\ell} \cdot \binom{a + b - 2u\ell}{a - u\ell}. \tag{5}$$

Corollary 1 ([7, Corollary 4]) *Let $u \geq 1$ and $S = \{(1, 0), (0, 1), (u, u + 1), (u + 1, u)\}$. Let $(a, b) \in \mathbb{N}^2$ with $a \leq b$, and write $a + b = (2u + 1)q + r$, with $0 \leq r \leq 2u$. Then,*

---

[4] They called it the Ciara feat. Missy Elliott example.

b / a	0	1	2	3	4	5	6	7	8	9	10	11
11	11	12	9	10	7	8	5	6	7	4	5	6
10	10	11	8	9	6	7	4	5	6	7	4	5
9	9	10	7	8	5	6	3	4	5	6	7	4
8	8	9	6	7	4	5	6	3	4	5	6	7
7	7	8	5	6	3	4	5	6	3	4	5	6
6	6	7	4	5	2	3	4	5	6	3	4	5
5	5	6	3	4	5	2	3	4	5	6	7	8
4	4	5	2	3	4	5	2	3	4	5	6	7
3	3	4	1	2	3	4	5	6	7	8	9	10
2	2	3	4	1	2	3	4	5	6	7	8	9
1	1	2	3	4	5	6	7	8	9	10	11	12
0	0	1	2	3	4	5	6	7	8	9	10	11

Fig. 4 Distances $d(a, b; S)$ for $0 \leq a, b \leq 11$ when $S = \{(1, 0), (0, 1), (2, 3), (3, 2)\}$

$$d(a, b; S) = \begin{cases} a + b - 2u \left\lfloor \frac{a}{u} \right\rfloor, & \text{if } a < uq; \\ a + b - 2uq, & \text{if } uq \leq a \leq \frac{a+b}{2}. \end{cases} \quad (6)$$

Something interesting and somewhat subtle happened here. Evoniuk et al. approached the problem of counting the number of minimal S-walks terminating at a point (a, b) by first determining the distance (remember that is the length of the shortest walk) from $(0, 0)$ to (a, b) and then counting walks of minimal length. Here, Iwanojko et al. first found a way to count the number of walks using ℓ long steps in Eq. (5) and then found the distance in Eq. (6) by determining the largest value of ℓ for which $\ell \leq a + b - 2u\ell$ and $a - u\ell \leq a + b - 2u\ell$, which would guarantee that the quantity on the right side of Eq. (5) is nonzero.

b/a	0	1	2	3	4	5	6	7	8	9	10
10	4	50	10	150	1215	101	1416	11046	546	7882	63056
9	1	16	130	20	255	1830	135	1740	12600	580	7882
8	6	3	36	250	31	355	2325	155	1860	12600	546
7	3	24	6	64	380	40	420	2520	155	1740	11046
6	1	9	48	10	88	460	44	420	2325	135	1416
5	3	2	15	72	12	96	460	40	355	1830	101
4	2	9	3	21	84	12	88	380	31	255	1215
3	1	4	12	4	21	72	10	64	250	20	150
2	1	1	4	12	3	15	48	6	36	130	10
1	1	2	1	4	9	2	9	24	3	16	50
0	1	1	1	1	2	3	1	3	6	1	4

Fig. 5 The number of minimal Q_3-walks terminating at each point (a, b) for $0 \leq a, b \leq 10$

Research Project 1 Explore the values of $d(a, b; S)$ and $|\mathscr{W}(a, b; S)|$ (or $|\mathscr{W}(a, b; S, \ell)|$) when $S = \{(1, 0), (0, 1), (u, v), (v, u)\}$ for general u and v.

4.2 Enumerating Walks with Steps of Fixed Length

For any $n \geq 2$, consider the set

$$Q_n = \{(1, 0), (0, 1)\} \cup \{(i, n - i) \ : \ 0 \leq i \leq n\}.$$

As before, we will say that the steps in the direction $(1, 0)$ and $(0, 1)$ are *short steps* and all other steps of the form $(i, n - i)$ are *long steps*.

For example, $Q_3 = \{(1, 0), (0, 1), (0, 3), (1, 2), (2, 1), (3, 0)\}$, and Fig. 5 shows data for the number of minimal Q_3-walks.

In Fig. 5, we observe an interesting phenomenon. If we fix a value $m = 3q + r$ with $0 \leq r < 3$ and consider all points (a, b) with $a + b = m$, then the number of minimal Q_3-walks terminating at (a, b) is a multiple of $\binom{q+r}{r}$. For example, when

$m = 7 = 3 \cdot 2 + 1$, the entries along the diagonal $(3, 9, 15, 21, 21, 15, 9, 3)$ are all divisible by $\binom{2+1}{1} = 3$.

Lemma 1 ([5, Lemma 2]) *Let $a, b \in \mathbb{N}$, and write $a + b = n \cdot q + r$ with $0 \leq r < n$. Then, any minimal Q_n-walk terminating at (a, b) uses exactly q long steps and r short steps. Consequently, $d(a, b; Q_n) = q + r$.*

Let $(a, b) \in \mathbb{N}^2$, and write $a + b = q \cdot n + r$ with $0 \leq r < n$. We can now use Lemma 1 to see why $|\mathscr{W}(a, b; Q_n)|$ is divisible by $\binom{q+r}{r}$. We can partition $\mathscr{W}(a, b; Q_n)$ into equivalence classes by declaring $\mathbf{s} \sim \mathbf{s}'$ if (1) $\mathbf{s}$ and $\mathbf{s}'$ use the same number of each step from Q_n and (2) the relative order of the long steps and the relative order of the short steps in $\mathbf{s}$ is the same as that in $\mathbf{s}'$. For example, in Q_3, the walks $(3, 0), (2, 1), (1, 0), (0, 1), (1, 2), (2, 1)$ and $(1, 0), (3, 0), (0, 1), (2, 1), (1, 2), (2, 1)$ are equivalent. The walks equivalent to $\mathbf{s}$ are determined by choosing r positions out of $q + r$ total steps in which we will place the (ordered list of) short steps.

Evoniuk et al. [5] left open the problem of determining the number of minimal length Q_n-walks terminating at a given point (a, b), as the linear algebra problem of determining all possible ways to write (a, b) as a sum of q long vectors and r short vectors seemed quite complicated. Iwanojko et al. [7] were able to solve this problem by framing it in the setting of the more general problem of counting walks by length.

For nonnegative integers, s, m, and p, define $\kappa(s, p, m)$ to be the number of ways to write $s = a_1 + \cdots + a_p$ such that $a_i \in \mathbb{N}$ and $a_i \leq m$ for all i. Formally, the quantity $\kappa(s, p, m)$ is called the number of *weak compositions* of s into p parts, each of which has a maximum value of m.

Theorem 5 ([7, Theorem 5]) *For any $n \geq 2$ and $a, b, \ell \in \mathbb{N}$,*

$$|\mathscr{W}(a, b; Q_n, \ell)| = \binom{a + b - (n-1)\ell}{\ell} \sum_{s=0}^{n\ell} \kappa(s, \ell, n)\binom{a + b - n\ell}{a - s}. \tag{7}$$

The formula in this theorem is quite complicated, but there are two instances in which it can be simplified. The simplification involves first being able to compute the values of $\kappa(s, \ell, n)$ when n is small and second using the *Vandermonde Identity* (which you proved in part (3) of Exercise 2) to collapse the sum on the right side of Eq. (7). Note that the summations in both of the following corollaries run from $i = 0$ to $i = \ell$ instead of from $i = 0$ to $i = n\ell$ as in Eq. (7).

Corollary 2 ([7, Corollary 6]) *For any $a, b, \ell \in \mathbb{N}$, the number of Q_2-walks terminating at (a, b) using ℓ long steps is*

$$|\mathscr{W}(a, b; Q_2, \ell)| = \binom{a + b - \ell}{\ell} \sum_{i=0}^{\ell} \binom{\ell}{i} \binom{a + b - \ell - i}{a - 2i}. \tag{8}$$

Corollary 3 ([7, Corollary 7]) *For any $a, b, \ell \in \mathbb{N}$, the number of Q_3-walks terminating at (a, b) using ℓ long steps is*

$$|\mathscr{W}(a, b; Q_3, \ell)| = \binom{a + b - 2\ell}{\ell} \sum_{i=0}^{\ell} \binom{\ell}{i} \binom{a + b - 2\ell}{a - 2i}. \tag{9}$$

5 An Interlude on Computational Techniques

You may have found yourself wondering how we generated the data in the previous sections. That data is important. The theorems we have proved would not have existed if we had not been able to start with data. One does not simply formulate these theorems; instead, it is necessary to start with data, observe patterns, and try to prove that those patterns are true in general.

We would be remiss if we did not mention the *Online Encyclopedia of Integer Sequences*, or *OEIS* [11]. Founded in 1964 by Neil Sloane, the *OEIS* is a database of integer sequences that is continuously updated by the researchers in the mathematical community. If you happen upon a sequence of integers arising from a combinatorial problem, the first thing you should do is check to see if it has an entry in the *OEIS*. It is a tremendous tool for finding connections to other known combinatorial results and also for making sure that someone else has not already proved the theorem you are about to discover. And if you find a sequence that is not yet in *OEIS*, be sure to add it!

It is worth pointing out that, like any new tool, the *OEIS* can be difficult to parse when you first encounter it. It takes time to learn to read the entries and understand all the tags associated with a given sequence. For mentors especially, it is worth taking time to introduce students to the different types of sequences that appear in the *OEIS*.

I have had one group of students work in Sage [15] to generate data, while other groups worked in Python. It is easy to write Python code in Sage—Sage just has more out-of-the-box mathematical libraries that are available for you to use. It would be quite difficult to make any progress on these problems without having a group member who has either taken some computer science classes or someone who is eager to learn how to do some basic coding. Nothing we present here requires terribly sophisticated programming techniques. I promise!

Writing your own code to generate data is a valuable learning experience, but we can talk about the general ideas that we used to generate our data. Let us start with the problem of determining the length of the shortest walk from $(0, 0)$ to (a, b)

using steps in a set S. As before, we will assume $(1, 0)$ and $(0, 1)$ belong to S so that it will always be possible to find *some* S-walk from any point $(0, 0)$ to (a, b), and hence there will always be a shortest such walk. This means all distances will be finite.

Let us go back to the distance data for $S = \{(1, 0), (0, 1), (2, 3), (3, 2)\}$ that is shown in Fig. 4. Our goal is to work recursively, just like we did when filling Pascal's triangle in Fig. 1. Let us imagine that we had partially filled in the values of $d(a, b; S)$ in Fig. 4 for all (a, b) with $a + b \leq 9$, and imagine we wanted to find the value of $d(4, 6)$.

How can we find a walk that terminates at $(4, 6)$? As an analogy to the block-walking argument we discussed when proving Theorem 1, we can focus on the last step in a walk terminating at $(4, 6)$ and look at all the points from which $(4, 6)$ can be reached in one step. To find these points, we subtract each vector in S from our target point, $(4, 6)$, and look at the (known!) distances to those points. This is illustrated in Fig. 6.

Here, the four points that are one step "behind" $(4, 6)$ are

$$(4, 6) - (1, 0) = (3, 6) \qquad d(3, 6; S) = 5$$
$$(4, 6) - (0, 1) = (3, 5) \qquad d(3, 5; S) = 5$$
$$(4, 6) - (2, 3) = (2, 3) \qquad d(2, 3; S) = 1$$
$$(4, 6) - (3, 2) = (1, 4) \qquad d(1, 4; S) = 5.$$

We conclude that a walk to $(4, 6)$ via $(3, 6)$ will have length 6, a walk to $(4, 6)$ via $(3, 5)$ will have length 6, a walk to $(4, 6)$ via $(2, 3)$ will have length 2, and a walk to $(4, 6)$ via $(1, 4)$ will have length 6. We pick the smallest number here to conclude that $d(4, 6; S) = 2$.

This example illustrates the general procedure for finding $d(a, b; S)$ for any set S. We must begin with one initial piece of knowledge: the distance from $(0, 0)$ to $(0, 0)$ is 0. Next, if you want to find the distance to a point $(a, b) \neq (0, 0)$, you could start by looking at all the points that can reach (a, b) in one step. In other words, look at all points in $\{(a, b) - s : s \in S\}$.

Here, we need to be a little careful to make sure we only consider points with nonnegative coordinates. My students have referred to this as "making sure we don't fall off the end of the earth. " For example, if we had wanted to find $d(1, 9; S)$ in Fig. 6, then $(1, 9) - (2, 3) = (-1, 6)$, which is not a point we can reach using steps in S. That means it is not feasible to consider this point.

Once we have found all the feasible points in $\{(a, b) - s : s \in S\}$, we pick the one whose distance is smallest and add 1 (why do we add 1?) to get $d(a, b; S)$.

Pseudocode that illustrates this procedure is shown in Algorithm 1.

Let us take a brief aside—this is not the first time we have viewed the value of $a + b$ as an important statistic in counting lattice walks. There are a few reasons why this value is important. First, it gives a sort of distance, or metric, associated with each point. When $S = \{(0, 1), (1, 0)\}$, the value $a + b$ tells us how many steps

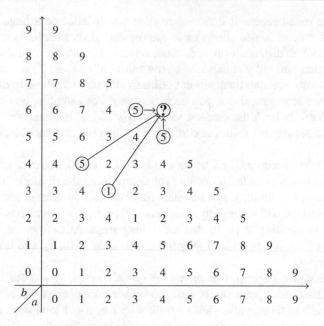

Fig. 6 Finding $d(4, 6; S)$ using recursion

Algorithm 1 Recursively find $d(a, b; S)$

Input A set S and a bound n
Output A table of values $d(a, b; S)$ for all $0 \leq a + b \leq n$.

  Initialize $d(0, 0; S) = 0$
  for $1 \leq k \leq n$ **do**
    for $0 \leq a \leq k$ **do**
      $b = k - a$
      for $s \in S$ **do**
        Set $(a', b') = (a, b) - s$.
        Check whether $a' \geq 0$ and $b' \geq 0$.
        If they are, store $d(a', b'; S)$
      end for
      Find the smallest value among all stored $d(a', b'; S)$. Call that distance D.
      Set $d(a, b; S) = D + 1$.
    end for
  end for

---

it takes to reach (a, b). That, in a sense, tells us how far (a, b) is from $(0, 0)$ in the context of this problem. A second reason why this value is important is that it strictly increases as we take steps. For example, if $(a, b) - s = (a', b')$ for some $s \in S$, then $a' + b' < a + b$. This means we can use $a + b$ as a quantity to induct upon in inductive proofs or recursive algorithms.

There are a few technical questions to address here. Perhaps a first question is "how do you store the values of $d(a, b; S)$?" We used dictionaries (or hash tables) to store our data because it does not require one to allocate a huge amount of memory all at once. It also allows us to use the coordinates (a, b) as keys that can be used to lookup distance values. Second, anyone who has taken a few computer science courses will tell you that storing the values of $d(a', b'; S)$ and then finding the smallest value among them are not efficient. If you know how to do this more efficiently, we recommend that you do it that way. For example, one can create a dummy variable to track the smallest value of $d(a', b'; S)$ that has been seen so far and compare subsequent values of $d(a', b'; S)$ to that dummy variable, updating as necessary.

The procedure for counting S-walks terminating at (a, b), and even tracking their length, is very similar to the procedure for determining the distance. To make the notation easier, we will once again assume that our set S consists of short steps and long steps, and we will recursively determine $|\mathscr{W}(a, b; S, \ell)|$, which is the number of S-walks terminating at (a, b) that use ℓ long steps. As before, the only initial value is that there is one walk of length 0 terminating at $(0, 0)$, and it uses 0 long steps.

In the recursive step, we find all values $(a', b') = (a, b) - s$ for $s \in S$ with $a', b' \geq 0$. Each such point can be reached by some number of walks with ℓ long steps, and each of those walks yields a walk with ℓ or $\ell + 1$ long steps terminating at (a, b). Pseudocode for this algorithm is given in Algorithm 2.

For this algorithm, we use nested dictionaries to store our information. We have a first dictionary whose keys are points (a, b) and whose values are also dictionaries. The dictionary linked to point (a, b) has keys that are the number of long steps in a walk and values that are the number of walks using that prescribed number of long steps terminating at (a, b).

Algorithm 2 Recursively find $|\mathscr{W}(a, b; S, \ell)|$

Input A set S and a bound n
Output A dictionary of values $|\mathscr{W}(a, b; S, \ell)|$ for all $0 \leq a + b \leq n$.
  Initialize $|\mathscr{W}(0, 0; S, 0)| = 1$
  for $1 \leq k \leq n$ **do**
    for $0 \leq a \leq k$ **do**
      $b = k - a$
      for $s \in S$ **do**
        Set $(a', b') = (a, b) - s$.
        Check whether $a' \geq 0$ and $b' \geq 0$.
        for ℓ such that $|\mathscr{W}(a', b'; S, \ell)| \neq 0$ **do**
          if s is a short step **then**
            add $|\mathscr{W}(a', b'; S, \ell)|$ to $|\mathscr{W}(a, b; S, \ell)|$.
          else if s is a long step **then**
            add $|\mathscr{W}(a', b'; S, \ell)|$ to $|\mathscr{W}(a, b; S, \ell + 1)|$.
          end if
        end for
      end for
    end for
  end for

Exercise 6 Prove that if $\mathbf{s} = s_1, s_2, \ldots, s_L$ is a walk of minimal length that terminates at (a, b), then $\mathbf{s}' = s_1, s_2, \ldots, s_{L-1}$ is a walk of minimal length that terminates at $(a', b') = (a, b) - s_L$. Why is this fact important for Algorithm 1?

Exercise 7 Write code to implement Algorithm 1 in your favorite programming language. Use it to recreate the data shown on the left of Fig. 2.

Exercise 8 Write code to implement Algorithm 2 in your favorite programming language. Use it to recreate the data shown on the right of Fig. 2. Then, verify that your data match with the result in Exercise 4. (Hint: how many long steps does a minimal path use?)

Exercise 9 How would you modify Algorithms 1 and 2 if you wanted to study a set S where some points (a, b) could not be reached by an S-walk?

6   Open Problems for Lattice Walk Enumeration

Aside from the results summarized in Sect. 4, there does not seem to be much that is known about this variation of the problem of enumerating lattice walks. There is tremendous space for further investigations. We use $\mathbb{Z}^2$ to denote the set of vectors in the xy-plane with integer coordinates.

> **Research Project 2** Pick an interesting set of vectors $S \subseteq \mathbb{N}^2$ or $S \subseteq \mathbb{Z}^2$ that has not already been discussed in Sect. 4. What can you say about the distances $d(a, b; S)$ or the number of walks $|\mathcal{W}(a, b; S)|$?

We saw in part (1) of Exercise 2 that the sum of the entries across a row of Pascal's triangle (or, as we have written it, a diagonal of Pascal's triangle) is $\binom{n}{0} + \binom{n}{1} + \cdots + \binom{n}{n} = 2^n$. How does this generalize for other sets?

> **Research Project 3** Pick an interesting set of vectors $S \subseteq \mathbb{N}^2$. Does anything interesting happen when you add
> $$\sum_{\substack{(a,b)\in\mathbb{N}^2 \\ a+b=n}} |\mathcal{W}(a, b; S)|?$$

(continued)

Here, we could ask about adding the number of minimal walks, walks of a fixed length, or total walks over all points (a, b) with a fixed coordinate sum.

Of course, there is also no reason to restrict oneself to studying walks in $\mathbb{Z}^2$. It follows from Theorem 2 that if one takes S to be the set of standard basis vectors in $\mathbb{R}^d$, then the number of S-walks terminating at $(a_1, \ldots, a_d)$ is given by the multinomial coefficient $\binom{a_1 + \cdots + a_d}{a_1, \ldots, a_d}$. As far as we know, nothing else is known.

Research Project 4 Pick an interesting set of vectors $S \subseteq \mathbb{N}^d$. What can you say about the distances $d(a_1, \ldots, a_d; S)$ or the number of walks $|\mathcal{W}(a_1, \ldots, a_d; S, \ell)|$ using a given number of long steps?

6.1   Generating Functions

Warning: The material presented in this section is substantially more difficult than everything we have done so far. Our presentation is very terse, which is not meant to imply that generating functions can be learned and mastered in an afternoon. Generating functions are a useful tool that take time to internalize, and we cannot present everything there is to learn about them here.

Generating functions require some intuition of power series from calculus. If you have not reached that part of your mathematical journey yet, you may want to skip this section.

In addition to—or in some cases, in lieu of—a combinatorial formula for the number of S-walks terminating at a point (a, b), an equally interesting solution to a combinatorial problem can be given through a *generating function*. Given a sequence of numbers $c_0, c_1, c_2, c_3, \ldots$, its generating function is the infinite series

$$f(x) = \sum_{n=0}^{\infty} c_n x^n.$$

Intuitively, a generating function can be viewed as one would think of a Taylor series in calculus, where you have a sort of "infinite polynomial" with meaningful coefficients. Unlike Taylor series, we do not need to worry about convergence properties of the function $f(x)$ because we will not be plugging in values for x. We simply view the variable x as a placeholder to display our coefficients for the world to see.

The main reason we use generating functions is that they afford us a whole wealth of algebraic manipulations that can be used to find unexpected connections between combinatorial sequences. One of the most such notable manipulations comes from the generating function for the sequence of Fibonacci numbers, which are defined by $F_0 = 0$, $F_1 = 1$, and $F_n = F_{n-1} + F_{n-2}$ for $n \geq 2$. Here, we have chosen to leave out *a lot*[5] of details in our calculations in the interest of highlighting a general technique (manipulating generating functions) and a cool result (the Fibonacci numbers are related to the golden ratio!). For more details and examples, especially, if you intend to use generating functions as a tool, it is worth reading Chapters 6 and 7 of Tucker's book [16] and working through some of the exercises there. Wilf's *Generatingfunctionology* [17] is also an excellent resource.

The generating function for the Fibonacci numbers is

$$F(x) = 0 + 1x + 1x^2 + 2x^3 + 3x^4 + 5x^5 + \cdots = \frac{-x}{x^2 + x - 1}. \qquad (10)$$

Exercise 10 The statement in Eq. (10) is not obvious. Prove it. Start by using the Fibonacci recurrence relation to prove that $(x^2 + x - 1) \cdot F(x) = -x$.

Next, we can[6] use partial fractions to rewrite

$$\frac{-x}{x^2 + x - 1} = \frac{A}{x - r_1} + \frac{B}{x - r_2},$$

where r_1 and r_2 are the roots of the quadratic $x^2 + x - 1$ and A and B are some numbers. Finally, we can write each term on the right as a geometric series. For example,

$$\frac{A}{x - r_1} = \frac{-A/r_1}{1 - x/r_1} = \frac{-A}{r_1} \cdot \sum_{n=0}^{\infty} \left(\frac{x}{r_1}\right)^n.$$

After a bit[7] more algebra (and resolving the constants[8] A and B), we arrive at a new expression

---

[5] Seriously, a lot. Perhaps a full page of algebra. Maybe even two.

[6] Let us not worry about doing this yet. Instead, we will take an elevator to the top of our mathematical ivory tower, where we will sit comfortably wrapped in the smug satisfaction that we *could* find all these values if we really had to.

[7] That may be an understatement.

[8] Smug satisfaction will not save us here. We actually need to do the work.

$$F(x) = \sum_{n=0}^{\infty} \frac{1}{\sqrt{5}} \left[\left(\frac{1+\sqrt{5}}{2} \right)^n - \left(\frac{1-\sqrt{5}}{2} \right)^n \right] x^n.$$

By comparing coefficients on both sides of this equation,[9] we obtain the surprising result that the nth Fibonacci number can be written as

$$F_n = \frac{1}{\sqrt{5}} \left[\left(\frac{1+\sqrt{5}}{2} \right)^n - \left(\frac{1-\sqrt{5}}{2} \right)^n \right].$$

When constructing a generating function for a sequence that has two variables, such as the sequence of distances $d(a, b; S)$ as a and b range over $\mathbb{N}^2$, we can add a second variable to get a bivariate generating function of the form

$$g(x, y) = \sum_{a=0}^{\infty} \sum_{b=0}^{\infty} d(a, b; S) x^a y^b.$$

Evoniuk et al. found a generating function for the number of Q_n walks.

Theorem 6 ([5, Theorem 3]) *For all $n \geq 2$, the number of minimal Q_n-walks can be computed by the generating function*

$$\sum_{(a,b)\in\mathbb{N}^2} |\mathscr{W}(a, b; Q_n)| x^a y^b = \sum_{q=0}^{\infty} \sum_{r=0}^{n-1} \binom{q+r}{r} \left(\sum_{i=0}^{n} x^i y^{n-i} \right)^q (x + y)^r. \quad (11)$$

This generating function carries with it the added information that $|\mathscr{W}(a, b; Q_n)|$ is divisible by $\binom{q+r}{r}$, where $a + b = q \cdot n + r$, which we previously observed. Aside from this result, very little seems to be known.

Research Project 5 Investigate the generating functions for the sequences $d(a, b; S)$ or $|\mathscr{W}(a, b; S)|$ as a and b range over $\mathbb{N}^2$.

The generating functions we have mentioned here are called *ordinary generating functions*. There is a related family of generating functions called *exponential generating functions*, which may be useful in exploring these problems.

---

[9] This part is important. Make sure you understand it.

Fig. 7 The area below the
lattice walk
$P = NNENEEEN$

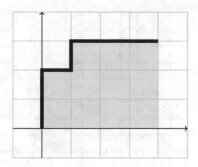

6.2 Counting According to a Statistic

Another way generating functions are used in combinatorics is to count lattice paths
according to some statistic. This section will be intentionally brief, with references
for further reading added as necessary.

Suppose we are given an N/E lattice walk from $(0, 0)$ to a point (a, b). We
can consider the *area* under the path as we would in a class on integral calculus.
However, because the path follows grid lines, we do not need antiderivatives to
compute the area. We simply need to count the number of boxes below the path and
above the x-axis. If P is a lattice walk, we use $\alpha(P)$ to denote its area. For example,
if $P = NNENEEE$, then $\alpha(P) = 11$, as illustrated in Fig. 7.

From this, we can define an *area generating function*

$$A(a, b; q) = \sum_{P} q^{\alpha(P)},$$

where P ranges over all N/E lattice walks from $(0, 0)$ to (a, b) and q is a dummy
variable. In other words, the coefficients in $A(a, b; q)$ count the number of lattice
walks terminating at (a, b) with a fixed area. For this reason, we think of $A(a, b; q)$
as a generating function that enumerates lattice paths *according to the area statistic*.
For example, Fig. 8 shows the ten N/E lattice walks from $(0, 0)$ to $(3, 2)$ with their
corresponding areas. From this, we obtain

$$A(3, 2; q) = 1 + q + 2q^2 + 2q^3 + 2q^4 + q^5 + q^6.$$

Exercise 11 Then find the area generating function $A(4, 3; q)$.

One notable feature of the area generating function is that evaluating it at $q = 1$
gives

$$\sum_{P} q^{\alpha(P)} \bigg|_{q=1} = \sum_{P} 1 = \binom{a + b}{a},$$

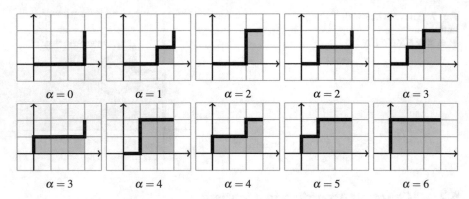

Fig. 8 Areas of N/E lattice walks terminating at $(3, 2)$

which just counts the number of lattice walks from $(0, 0)$ to (a, b). Because of this, we say $A(a, b; q)$ is a *q-analog* of the binomial coefficient $\binom{a+b}{a}$. The interesting fact is that we can write $A(a, b; q)$ as a polynomial in q without having to compute the areas of all the lattice walks from $(0, 0)$ to (a, b) first. We refer to Stanley's *Enumerative Combinatorics* [13] for more details on q-analogs of binomial coefficients and Krattenthaler's survey on lattice path enumeration [8, Section 10.19] for a discussion of their relation to N/E lattice walks.

However, there is no reason to restrict our attention to N/E lattice walks here. An area generating function $A(a, b; q, S)$ can be defined analogously for any set S of allowable steps.

Research Project 6 Explore the area generating function $A(a, b; q, S)$ for other sets of allowed steps.

Research Project 7 Explore generating functions for S-walks based on combinatorial statistics other than area.

The area generating function $A(a, b; q, S)$ when $S = \{(1, 0), (0, 1), (1, 1)\}$ (i.e., for N/E/D lattice walks) was investigated by Carlos et al. [3], who also gave a combinatorial interpretation of $A(a, b; q, S)$ evaluated at $q = -1$ in this special case. They showed that the evaluation of $A(a, b; q, S)|_{q=-1}$ is, up to a predictable multiple by 1 or $\sqrt{-1}$, the number of *palindromic* N/E/D lattice walks from $(0, 0)$ to (a, b).

7   Counting Walks with Restrictions

Another classical question in enumerative combinatorics asks how many N/E lattice walks in the plane start at $(0, 0)$, end at (a, a), and never cross above the line $y = x$. For example, when $a = 3$, there are five such walks, which are illustrated in Fig. 9.

The numbers we get here, $1, 2, 5, 14, 42, \ldots$, are called the *Catalan numbers*, and they are ubiquitous in combinatorics. In fact, Richard Stanley has collected a volume of different interpretations of the Catalan numbers [14], which contains over 200 different interpretations.

This leads to a natural generalization, which was defined by Evoniuk et al.
Definition 3 Let $S \subseteq \mathbb{N}^2$. An *S-Catalan walk* is an *S*-walk that never crosses above the line $y = x$. For integers a and b, use $C(a, b; S)$ to denote the set of *S*-Catalan walks terminating at (a, b).

Therefore, for $S = \{(1, 0), (0, 1)\}$, the number of *S*-Catalan walks terminating at (a, a) is the ath Catalan number. More generally, the number of *S*-Catalan walks terminating at (a, b) is given by the formula $\frac{a-b+1}{a+b+1}\binom{a+b+1}{a+1}$. We refer to Krattenthaler [8] for a proof of this fact, along with a number of other interesting results on lattice walks and lattice walks with geometric restrictions.

Hilton and Pedersen [6] studied another instance of this problem, enumerating *m-Dyck paths*, which are walks from $(0, 0)$ to $(mn, 0)$ that use steps $(1, 1)$ or $(1, 1-m)$ and do not pass below the x-axis. If we rotate the plane, we can see that this can also be translated into a problem about *S*-walks that do not cross above or below the line $y = x$. These paths were further generalized by Song [12], who considered walks from $(0, 0)$ to (mn, n) using north, east, and diagonal steps that do not cross below the line $x = my$.

Evoniuk et al. [5, Section 6] explored *S*-Catalan walks for sets of the form $S_n :=\{(i, n - i) \ : \ 0 \leq i \leq n\}$. Because each vector $(a, b) \in S_n$ has coordinate sum $a + b = n$, all walks terminating at a point (a, b) have the same length, and only points whose coordinate sum is divisible by n can be reached.

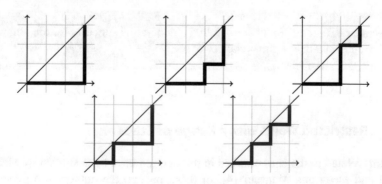

Fig. 9 N/E lattice walks from $(0, 0)$ to $(3, 3)$ that do not cross below $y = x$

b\a	0	1	2	3	4	5	6	7	8	9	10	11	12	13	14	15	16	17	18
9										120									
8									0	0	184								
7								0	52	0	0	234							
6							13	0	0	68	0	0	212						
5						0	0	18	0	0	64	0	0	158					
4					0	6	0	0	21	0	0	50	0	0	99				
3				2	0	0	7	0	0	16	0	0	30	0	0	50			
2			0	0	2	0	0	5	0	0	9	0	0	14	0	0	20		
1		0	1	0	0	2	0	0	3	0	0	4	0	0	5	0	0	6	
0	1	0	0	1	0	0	1	0	0	1	0	0	1	0	0	1	0	0	1

Fig. 10 Number of S_3-Catalan walks terminating at points (a, b) with $0 \le a + b \le 18$

For $S_2 = \{(2, 0), (1, 1), (0, 2)\}$, the number of S_2-Catalan walks terminating at the point (a, a) is the ath *Motzkin number*, which is another famous sequence in enumerative combinatorics.

Moving on to $n = 3$, the number of S_3-Catalan walks terminating at a point (a, b) is shown in Fig. 10.

The number of S_3-Catalan walks terminating at (a, a) with a divisible by 3 appears as *OEIS* entry A292437.

> **Research Project 8** What is the generating function for the number of S_n-Catalan walks terminating at (a, a)? How about at (a, b)? Is there a closed formula for the number of such walks?

> **Research Project 9** Enumerate or find a generating function for the number of S-Catalan walks for other choices of S.

7.1 Restricted Walks with a Range of Steps

Problems related to the ones outlined in this section have been studied by Mansour [9, 10] and Deng and Mansour [4]. In these papers, the authors use generating functions to explore walks that stay in a fixed region of the plane, but with greater flexibility in the set of allowable steps.

Mansour [9] began by exploring the following problem. Let $\mathscr{A} = \{(a, b) \in \mathbb{Z}^2 : 0 \leq b \leq a\}$, that is, the points in the first quadrant that lie above the line $y = x$. As a step, one is allowed to jump from a point $(i, j) \in \mathscr{A}$ to a point $(i', j + 1)$ with $i \leq i' \leq j + 1$. In how many ways can walk from the origin to a point $(a, b) \in \mathscr{A}$? In this problem, the set of allowable steps varies depending on the current location of the walk—from the point (i, j), the allowable steps are $\{(m, 1) : 0 \leq m \leq j + 1 - i\}$. It is important to note that the walk must always remain in the region $\mathscr{A}$.

Later, Mansour [10] proposed a variation on the above problem, studying walks that remain in the same region $\mathscr{A}$, but in which the allowable steps from a point (i, j) are $\{(m, 1) : 0 \leq m \leq j + 1 - i\} \cup \{(0, 2)\}$. Mansour showed that the number of such walks is in bijective correspondence with certain types of even trees.

Another generalization, studied by Deng and Mansour [4, Problem 1], can be summarized as follows: let $k \geq 2$ and let $\mathscr{A}_k = \{(a, b) \in \mathbb{Z}^2 : 0 \leq a \leq (k - 1)b\}$; that is, the integer points in the region of the first quadrant bounded by the y-axis and the line $y = \frac{1}{k-1}x$. As a step, you are allowed to move from a point $(i, j) \in \mathscr{A}_k$ to a point $(i', j + 1) \in \mathscr{A}_k$ with $i \leq i'$. What can be said about the numbers $a_k(i, n)$, which is the number of ways to walk from $(0, 0)$ to the point (i, n) for $0 \leq i \leq (k - 1)n$?

In this problem, we see that the set of allowable steps is $S = \{(m, 1) : m \geq 0\}$, which is an infinite set, but the requirement that the walk stays within the region $\mathscr{A}_k$ effectively makes S finite for any fixed target point (i, n).

As a further variation, Deng and Mansour [4, Problem 2] explore walks that stay in the region above the line $y = x$ in the first quadrant with allowable steps $S = \{(m, 1) : m \geq 0\} \cup \{(1, 0)\}$.

From these problems, we have access to a seemingly infinite number of questions related to S-walks and generating functions.

Research Project 10 Let $k, \ell \in \mathbb{N}$ with $k < \ell$, and fix a set S of allowable steps. Let $\mathscr{A}_{k,\ell} = \{(a, b) \in \mathbb{N}^2 : ka \leq b \leq \ell a\}$. How many S-walks from $(0, 0)$ to (a, b) stay in the region $\mathscr{A}_{k,\ell}$?

Finally, Bonin et al. [2] studied generating functions (in the spirit of Sect. 6.2) associated with various combinatorial statistics for N/E/D lattice walks that stay weakly below the line $y = x$, which are also called *Schröder paths*.

8 Conclusion

We hope that the problems outlined in this chapter can be used to inspire new undergraduate research projects. The combinations of questions that can be asked and sets of allowable step vectors that can be chosen is seemingly endless. Some of those questions will be approachable, while others may be very difficult. We encourage you to ask a lot of questions, gather data, find patterns, and report on those patterns that can be explained.

References

1. Benjamin, A.T., Quinn, J.J.: Proofs that really count, *The Dolciani Mathematical Expositions*, vol. 27. Mathematical Association of America, Washington, DC (2003). The art of combinatorial proof
2. Bonin, J., Shapiro, L., Simion, R.: Some q-analogues of the Schröder numbers arising from combinatorial statistics on lattice paths. J. Statist. Plann. Inference **34**(1), 35–55 (1993). https://doi.org/10.1016/0378-3758(93)90032-2. URL https://doi-org.offcampus.lib.washington.edu/10.1016/0378-3758(93)90032-2
3. Carlos, E., Klee, S., Pham, H., Shrock, B.: Area statistics for Delannoy paths (2021). In preparation.
4. Deng, E.Y.P., Mansour, T.: Three Hoppy path problems and ternary paths. Discrete Appl. Math. **156**(5), 770–779 (2008). https://doi.org/10.1016/j.dam.2007.08.015
5. Evoniuk, J., Klee, S., Magnan, V.: Enumerating minimal length lattice paths. J. Integer Seq. **21**(3), Art. 18.3.6, 12 (2018)
6. Hilton, P., Pedersen, J.: Catalan numbers, their generalization, and their uses. Math. Intelligencer **13**(2), 64–75 (1991). https://doi.org/10.1007/BF03024089
7. Iwanojko, N., Klee, S., Lasher, B., Volpi, E.: Enumerating lattice walks with prescribed steps. J. Integer Seq. **23**(34), Art. 20.4.3, 15 (2020)
8. Krattenthaler, C.: Lattice path enumeration. In: Handbook of enumerative combinatorics, Discrete Math. Appl. (Boca Raton), pp. 589–678. CRC Press, Boca Raton, FL (2015)
9. Mansour, T.: Combinatorial methods and recurrence relations with two indices. J. Difference Equ. Appl. **12**(6), 555–563 (2006). https://doi.org/10.1080/10236190600637767
10. Mansour, T.: Recurrence relations with two indices and even trees. J. Difference Equ. Appl. **13**(1), 47–61 (2007). https://doi.org/10.1080/10236190601069333
11. Sloane, N.J.A.: (2019). The On-Line Encyclopedia of Integer Sequences, http://oeis.org
12. Song, C.: The generalized Schröder theory. Electron. J. Combin. **12**, Research Paper 53, 10 (2005)
13. Stanley, R.P.: Enumerative combinatorics. Volume 1, *Cambridge Studies in Advanced Mathematics*, vol. 49, second edn. Cambridge University Press, Cambridge (2012)
14. Stanley, R.P.: Catalan numbers. Cambridge University Press, New York (2015). https://doi.org/10.1017/CBO9781139871495
15. Stein, W., et al.: Sage Mathematics Software (Version 8.8). The Sage Development Team (2019). http://www.sagemath.org
16. Tucker, A.: Applied combinatorics, sixth edn. John Wiley & Sons, Inc., New York (2012)
17. Wilf, H.S.: generatingfunctionology, third edn. A K Peters, Ltd., Wellesley, MA (2006)

The Mathematics of Host-Parasitoid Population Dynamics

Brooks Emerick

Abstract

Host-parasitoid interactions make up an important class of consumer resource dynamics. A parasitoid is an organism that spends most of its life-cycle attached to or inside the host. Unlike an actual parasite, the parasitoid eventually kills the host. During a particular season of the year, known as the vulnerable period, the parasitoid injects eggs into the host larvae. Parasitoid larvae, then, emerge from the host, effectively killing it. Discrete-time models are usually employed to simulate host-parasitoid interactions because reproduction occurs at the same fixed date every year. Early models of this phenomenon include the discrete-time Nicholson–Bailey model, which is known to be unstable, i.e., coexistence is impossible. More recent models take advantage of a hybrid technique that uses continuous dynamics to model the vulnerable period so that stabilizing characteristics can be mechanistically incorporated. We take the reader on a tour of basic discrete models, dynamic models of the vulnerable period, and an introduction to the semi-discrete framework. Along the way, we demonstrate the model building process, develop, and solve systems of differential equations, and perform linear stability analyses. Ultimately, readers will feel comfortable with the host-parasitoid modeling process and will be able to explore sophisticated modeling scenarios, potentially discovering new models that provide biological insight into the complex dynamics of the host-parasitoid interaction.

B. Emerick (✉)
Kutztown University, Kutztown, PA, USA
e-mail: bemerick@kutztown.edu

© The Author(s), under exclusive license to Springer Nature Switzerland AG 2022
E. E. Goldwyn et al. (eds.), *Mathematics Research for the Beginning Student,*
Volume 2. Foundations for Undergraduate Research in Mathematics,
https://doi.org/10.1007/978-3-031-08564-2_7

Suggested Prerequisites *Knowledge of basic calculus concepts, and the ability to manipulate multivariable functions and compute partial derivatives are all essential. Some acquaintance with solving linear first-order differential equations, separation of variables, and an acquaintance with trace and determinant of matrices will be helpful. Experience with coding in MATLAB is recommended, but not required.*

1   Introduction

We start our study of the mathematics of the host-parasitoid interaction by introducing various parasitoid species. It is important (and interesting) to know the biology behind what we are actually modeling. Further, it is quite surprising to actually realize that there are real-life Xenomorphs from the 1979 hit movie *Alien*, but on a much smaller scale. In the sections below, we introduce this interesting parasitic species and the various behavioral patterns that we can simulate with mathematical modeling. Then, we introduce mathematical skills needed to analyze the models that we create throughout the chapter. The best part of modeling the host-parasitoid interaction is the versatility of the modeling process—a new model can be created with ease and the mathematical discoveries that result from it can be extremely interesting, if not enlightening and fun.

1.1   What Are Parasitoids?

The iconic "chestburster" scene of the original 1979 sci-fi classic movie *Alien* is the stuff of movie legend. In this scene, the late English actor John Hurt plays Kane, an executive officer among six other crew members of a spacecraft that is returning to Earth from a distant moon. A distress call from a nearby ship attracts the crew and they respond in an attempt to provide aid. Kane finds a group of egg-like pods on the ship and upon investigation, a creature leaps from the pod and attaches to his face. Kane is incapacitated for several days with the creature still attached until it eventually falls off and dies. Once he regains consciousness, Kane is seemingly fine and he sits down to eat with his fellow crew members. All seems well until he begins to convulse in pain. The other crew members attempt to restrain him when a small alien creature bursts violently through his chest, killing him, before escaping into the bowels of the spacecraft. Unbeknownst to Kane and the crew, the creature had infected him with an alien embryo while it was attached to his face. Kane's body was essentially used as an incubator for the second stage of the creature's life.

This is exactly how a parasitoid behaves, although the life-cycle of the Xenomorph (the alien from the movie) is somewhat different. The basic idea is that the "facehugger," which is the Stage 1 Xenomorph, has a single purpose: to implant the chestburster egg into an unwilling host. The host in this scenario is a human. Once the chestburster has been successfully implanted into its host, it later...how do I say this..."emerges." The human host does not survive and the alien life-form eventually undergoes a metamorphosis, growing into the Stage 3 adult, fully loaded with acidic blood and an inner pharyngeal jaw capable of extreme carnage. Not to spoil the entire movie, I will simply say that the creature terrorizes the remaining crew of the interstellar cruiser dubbed Nostromo. If you have not seen the movie and you would like to know how the story ends, please see Exercise 1.

The egg implantation and chestbursting characteristics of the Xenomorph are physiological behaviors of actual parasitoids, although it occurs on a much smaller scale. The movie presents a terrifying depiction of *oviposition*, the implantation of a viable parasitoid egg into the host. The parasitoid life strategy is common among many wasp species (Order Hymenoptera). It is estimated that more than 65,000 species of Hymenoptera exist in the world, most of which have not been formally described [31]. The parasitoid life-style also occurs in some fly species as well as in beetles, moths, and butterflies, albeit rarely [20]. All models that we'll consider are created with the parasitic wasp in mind.

A typical parasitic species will live off of its host, but never purposely kill it as in the case of the flea or tick. In contrast, the host of a parasitoid will certainly die when the parasitoid larvae eventually emerge from it. In a parasitic relationship, the parasite needs the host to complete its full life-cycle and this is also true for the parasitoid. In fact, the host body is the source of nourishment of young parasitoid larvae [28,40]. Immature parasitoids are blind, maggot-like, and soft-bodied in form and they remain inside the host for the duration of the larval stage. They will feed on the host body both internally and externally until they are ready to pupate [41]. The free-living adults are mobile, with wings used for dispersion and antennae for detecting a variety of chemical cues that are crucial for finding a mate and healthy hosts for development of their progeny [42]. Once it has mated, the female parasitoid has complete control over the fertilization of their eggs. Male larvae develop from unfertilized eggs and females develop from fertilized eggs. Thus, a mated female can control the male/female ratio of the progeny [15]. The egg is delivered by its *ovipositor*. The ovipositor is like a stinger, but elongated and tubular and designed specifically to deposit eggs into or onto the host [31]. The ovipositor is a unique tool. It is not only the primary mechanism for oviposition but it also works as a probe that detects mechanical and chemical stimuli from the host body to determine if it is healthy and viable [31,42]. The search efficacy and egg-laying proficiency of the female adult is essential for the survival of the species.

From a modeling perspective, we view the life-cycles of the host and the parasitoid as parallel structures. Each insect has a typical life-cycle broken into the following disjoint compartments: egg, larval, pupae, and adult (see Fig. 1). At some point during the year, the female parasitoid adults interact with the host in its larval stage [14]. This period during which oviposition occurs, will be referred to as the *vulnerable period*. Once the period is over, a proportion of the host population is effectively parasitized. This proportion will not survive to the next compartment of the host life-cycle. However, this same proportion will create a generation of new parasitoid pupae once they have successfully grown from the host's body. Oviposition is a simultaneous act of self-reproduction and host-suppression. This poses an interesting scenario: If the host species is over-suppressed, the parasitoid

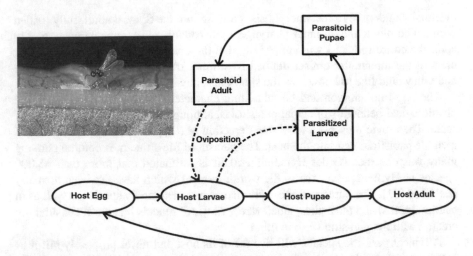

Fig. 1 Life-cycle of the host and the parasitoid. Inset shows the picture of a parasitoid wasp laying an egg into the body of its host (spotted alfalfa aphid). Picture taken with permission from https://extension.umd.edu/resource/aphid-parasitoids

will not be able to reproduce. This is typical of a predator-prey-type scenario where the prey is over-exploited until the predator population declines. The prey, then, are able to reproduce and grow, thus creating an abundant food resource for the predator. The predator over-exploits the prey and the cycle continues. We'll see that the act of locating a host, assessing a host, and successfully ovipositing into the host is key for the survival and persistence of both the host and parasitoid species.

Exercise 1 Watch the movie *Alien*.

1.2   Host and Parasitoid Behavior

Although the basic concept of the movie *Alien* captures true behavior of parasitic wasps, the movie got a lot of things wrong. I mean, what is with the acidic blood? In the movie, the parasitoid is the antagonist, but realistically, parasitoids are usually the good guys. The host species of some parasitoids are pests. I guess that means humans are the pests? Introducing a natural enemy to control the pest is a technique that has been employed since the beginnings of biological pest control [15]. Simulating the interaction between hosts and parasitoids can, therefore, provide insight and even confirm experimental data on the effectiveness of parasitoid species to control a pest population.

To control the pest population, the parasitoid is introduced to limit the spread and destruction of the host. It is important to note that the parasitoid typically does not eradicate the host population. Once introduced, the parasitoid and host coexist and over time will both persist in the environment [15]. Mathematical models are created to simulate the behavior of this interaction. Pest control scientists are interested in the behaviors of parasitoids that lead to more efficient pest eradication. They are also interested in predictive simulations that reflect biological data or lab experiments. Semi-discrete mathematical models, which are the main focus of this research project, take into account host and parasitic behaviors that can be mechanistically incorporated during the vulnerable period. This is in contrast to purely discrete models, which typically employ mechanisms that occur either before or after the interval in which the two populations coincide. From a modeler's perspective, incorporating various mechanisms into the model is the best part! What kind of behavior can we model and what type of results are there to discover?

As many mathematical models suggest, efficiency in the parasitoid's ability to locate and successfully oviposit determines if coexistence is possible [12,13,27,37]. This can happen in a variety of ways. One way is to consider alternative forms of host mortality. The parasitoid is not the only predator of the host, and so incorporating competition for the host resource is a stabilizing mechanism [3]. We explore a mathematical model that incorporates this competitive scenario in Sect. 3.2. Another interesting behavior of the parasitoid is host-feeding [19]. There exists a tendency of synovigenic parasitoids to consume or partially destroy the host without laying eggs inside. Generally, an adult female parasitoid emerges every year with less eggs than she can potentially oviposit in her lifespan. Therefore, a parasitoid will feed on hosts to gain the necessary energy to mature additional eggs [23, 24]. Although host-feeding is ultimately a necessary process for creating new eggs, it results in a loss of potential hosts used for oviposition. The act of host-feeding effectively works against long term persistence of the parasitoid species [6]. Both of these behaviors have been studied biologically and mathematically and an extension to existing models is discussed throughout the chapter.

A parasitoid species spends a proportion of the vulnerable period searching for a suitable host [27, 40]. The landscape in which a female parasitoid forages may be spatially heterogeneous in nature, consisting of several host larvae patches. Studies have shown that once a parasitoid is within a suitable patch, they may not leave. In fact, they will search until reaching the boundary and turn around to continue searching [1,26,30]. However, search efficiency is key for reproduction and parasitoids may cut their losses and flee to a different patch in search of a suitable host [4,14,40]. Theoretical dynamics of spatially distributed host-parasitoid systems have been extensively studied [16, 17]. In previous work, Emerick et. al. [13] study parasitoid patch-use in two locations using the semi-discrete framework. One interesting result is the idea of a host refuge, which essentially means a proportion of hosts continually escape parasitism every cycle. Again, this is a form of searching inefficiency that yields stability in the system [13, 22]. We explore this in Sect. 2.3. In general, a semi-discrete formulation of parasitoid migration across a broad network is a project of interest and is considered in Research Project 4.

Lastly, multiple species of hosts and parasitoids may interact in the same vulnerable period. Competition can lead to coexistence in host-parasitoid systems. Using the semi-discrete framework to model more than one parasitoid population that feeds on the same host species is an interesting project to consider (see Research Project 1 and 3). Another type of competition that has not been studied in the semi-discrete framework is the hyperparasitoid species. A hyperparasitoid is a type of parasitoid that attacks parasitoids developing inside the previously parasitized host [15]. In this scenario, the hyperparasitoid hijacks a parasitized host, thereby killing the parasitoid egg within and using it as a vehicle of its own reproduction. We suggest using the semi-discrete modeling approach to explore this new dynamic with previously studied methods of stabilization, including the host mortality or host refuge, in Research Project 2.

In the next section, we introduce the basic mathematical methods used to analyze discrete dynamical systems in one and two dimensions. This will give the reader the necessary tools to succeed in studying the semi-discrete models formulated for the host-parasitoid interaction. All MATLAB codes associated with the figures and models in the following sections can be found in the github repository [11].

1.3   Prerequisites

Often, the goal of modeling the host-parasitoid interaction is to determine if and how coexistence is possible. Given a specific set of assumptions and parameters, can we effectively describe how and why the host and parasitoid populations persist as time approaches infinity? This is usually the challenge presented for any ecological model. Luckily for us, the idea of coexistence corresponds directly to the mathematical definition of a nontrivial *equilibrium* or *fixed point*. A nontrivial fixed point is a fixed point that takes on a nonzero value.

We define a general discrete model in one-dimension using the following recursive relationship

$$x_{t+1} = f(x_t) \qquad x_0 = \text{given}, \tag{1}$$

where x_t represents the density of population x at time t. The variable t is discrete, taking on positive integer values that denote the current year. The initial condition, x_0, is known. Equation 1 is a general one-dimensional discrete dynamical system, where the population total in the next year is completely determined by the function f applied to the population total in the previous year. For any given initial condition, the result of Eq. 1 is an infinite sequence of values, where each value is computed by repeated application of the function f. This sequence is called a *trajectory*. We seek answers to questions relating to the convergence of this sequence. In other words, where do the values of this trajectory end up as $t \to \infty$? To this end, we study the fixed points of the system.

Definition 1 Given the system defined by Eq. 1, if $x^* = f(x^*)$, then x^* is a *fixed point*.

A fixed point never moves, hence its name. If we start at x^*, the system remains in a fixed state, staying at x^* at every iteration. That is, a trajectory with $x_0 = x^*$ is an infinite sequence of the repeated value x^*. From the definition, we see that x^* satisfies $x_{t+1} = x^* = x_t$, which shows that the next term is identical to the previous. Graphically, fixed points occur at the intersection of the line $y = x$ and the graph of $y = f(x)$. A discrete system may have several fixed points.

When analyzing a dynamical system, we first determine the system's fixed points because interesting behavior occurs "near" these special points. For example, if a trajectory starts at the fixed point, $x_0 = x^*$, it will never leave that point. However, if a trajectory contains a value that is near to this fixed point, we would like to answer the following question: do the remaining values of the trajectory approach the fixed point or move away from the fixed point? In this sense, we may classify fixed points as *attracting*, *repelling*, or *neutral*, which essentially describes the behavior of any given trajectory in the vicinity of that fixed point. The following theorem is used to determine linear stability of any fixed point to a one-dimensional discrete system.

Theorem 1 *Suppose x^* is a fixed point of the one-dimensional discrete dynamical system defined by Eq. 1. If $|f'(x^*)| < 1$, then x^* is linearly stable (attracting). If $|f'(x^*)| > 1$, then x^* is linearly unstable (repelling). If $|f'(x^*)| = 1$, then x^* is neutral or indifferent.*

A 1-D Example To demonstrate this theorem, we consider the classic Logistic Map with $r = 2$. Let $f(x) = 2x(1 - x)$, then our discrete system has the form

$$x_{t+1} = 2x_t(1 - x_t). \tag{2}$$

The Logistic Map is a one-dimensional population model developed by Robert May, where the quantity x_t is meant to represent the ratio of the existing population to the maximum possible population. Thus, x_t is a proportion. Performing a linear stability analysis, we substitute x^* for both x_{t+1} and x_t, and solve for x^*. This yields two fixed points $x_0^* = 0$ and $x_1^* = 1/2$. The derivative of f is $f'(x) = 2 - 4x$ (check for yourself). Applying the theorem, we obtain

$$x_0^* : \quad |f'(x_0^*)| = |f'(0)| = 2 > 1$$
$$x_1^* : \quad |f'(x_1^*)| = |f'(1/2)| = 0 < 1,$$

which means $x_0^* = 0$ is repelling and $x_1^* = 1/2$ is attracting. There exists a set of initial values x_0 such that $x_t \to 1/2$ as $t \to \infty$. This set of initial values that is attracted to x_1^* is called the *basin of attraction*. For this model, the basin of attraction for x_1^* is $(0, 1)$ as demonstrated in Fig. 2.

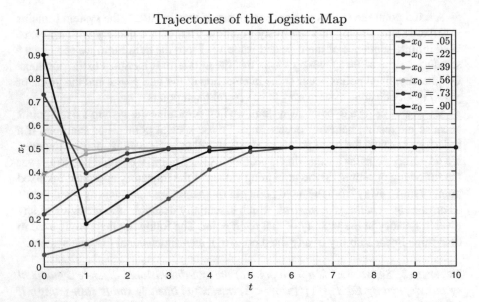

Fig. 2 Several trajectories of the Logistic Map with $r = 2$. We can see that the fixed point $x_1^* = 1/2$ is linearly stable since all trajectories with initial values not equal to $1/2$ eventually approach the value $1/2$ over time. See the github repository to view the MATLAB code used to produce this figure [11]

Exercise 2 Consider the general Logistic Map, $f(x) = rx(1 - x)$, where r is a parameter. Find the fixed points of this model and determine the conditions on r for which the nontrivial fixed point is stable. What happens when $r = 3$? (Hint: fixed points can sometimes depend on system parameters.)

It is helpful to study one-dimensional systems because the same ideas can be used to analyze higher dimensional systems. However, when we are dealing with more than one population (i.e., more than one equation), the generalization of the linear stability analysis can be complicated. In order to study two–dimensional systems, we must know how to compute fixed points, the *Jacobian matrix*, and the *eigenvalues* of a 2×2 matrix.

To introduce these mathematical ideas, we start with a general two-dimensional discrete system. We assume there are two interacting populations x_t and y_t whose next generation is determined by the multivariable functions f and g:

$$x_{t+1} = f(x_t, y_t) \qquad x_0 = \text{given} \tag{3}$$

$$y_{t+1} = g(x_t, y_t) \qquad y_0 = \text{given.} \tag{4}$$

A trajectory of this system is an infinite sequence of ordered pairs (x_t, y_t). Just as in the one-dimensional case, we seek the fixed points of this system.

Definition 2 Given the system defined by Eqs. 3 and 4, if (x^*, y^*) satisfies

$$x^* = f(x^*, y^*) \tag{5}$$

$$y^* = g(x^*, y^*), \tag{6}$$

then (x^*, y^*) is a fixed point.

Similar to the one-dimensional case, trajectories may be attracted to or repelled from a fixed point. We wish to study the stability of fixed points and in order to do this, we construct a 2×2 matrix of partial derivatives known as the Jacobian.

Definition 3 Consider the system given by Eqs. 3 and 4. The Jacobian matrix, denoted by $J(x, y)$ is the 2×2 matrix made up of the partial derivatives of $f(x, y)$ and $g(x, y)$, given by

$$J(x, y) = \begin{bmatrix} \frac{\partial f}{\partial x} & \frac{\partial f}{\partial y} \\ \frac{\partial g}{\partial x} & \frac{\partial g}{\partial y} \end{bmatrix}. \tag{7}$$

A function of two variables (like f or g in the case above), have two partial derivatives: a derivative with respect to x and a derivative with respect to y. Because our variable depends on both x and y, we can discuss the rate of change of our function in the direction of each independent variable. As in single variable calculus, given a function $f(x)$, the notation df/dx represents the instantaneous rate of change of the function values with respect to the variable x. With a multivariable function, $f = f(x, y)$, the notation $\partial f/\partial x$ represents the instantaneous rate of change of f with respect to the variable x and $\partial f/\partial y$ represents the rate of change with respect to y. Graphically, a partial derivative tells us the slope or shape of the function in the direction of a particular variable. We can use the rules of introductory calculus to compute partial derivatives. Consider the example below.

Partial Derivative Example Let f be a multivariable function defined by

$$f(x, y) = 5x^2 + e^y - x^3 y^2 + \cos(xy).$$

When differentiating with respect to x, the rules of differentiation apply in the usual way but now you treat the variable y as if it were a constant (since we are not considering change in that variable). When differentiating with respect y, you treat x as a constant. The partial derivatives of f are

$$\frac{\partial f}{\partial x} = 10x - 3x^2 y^2 - y \sin(xy)$$

$$\frac{\partial f}{\partial y} = e^y - 2x^3 y - x \sin(xy).$$

Exercise 3 Determine all partial derivatives of the following multivariable function:

$$f(x, y, z) = xy \left(5 \sin^2(z) + 1\right) + \frac{e^{y^2+1}}{xz}.$$

Recall in the one-dimensional case that linear stability is determined by the magnitude of the derivative of f evaluated at the fixed point x^*. We evaluate the Jacobian at the fixed point (x^*, y^*), which yields a 2×2 matrix, denoted simply as J^*. How do we determine the "magnitude" of this matrix evaluated at the fixed point? We determine the eigenvalues of the matrix. The eigenvalues of this matrix provide insight into the stability of the system. For a general explanation of the interpretation of eigenvalues and how to compute them, we refer the interested reader to [39].

All of the models we consider are two-dimensional. The eigenvalues of J^*, denoted by λ, can be found by computing the zeros of the following characteristic polynomial:

$$\lambda^2 - \text{tr}(J^*)\lambda + \det(J^*), \tag{8}$$

where $\text{tr}(J^*)$ is the *trace* of J^* and $\det(J^*)$ is the *determinant* of J^*. The trace and determinant are two characteristics of any $n \times n$ matrix. The trace is the sum of the diagonal elements of a matrix and the determinant is a value that can determine certain matrix properties such as invertibility. The determinant can be computed using several methods, one of which is called the method of cofactor expansion [39]. This method is typically used for large square matrices. For 2×2 matrices, the determinant can be found using the following calculation:

$$A = \begin{bmatrix} a & b \\ c & d \end{bmatrix} \quad \Rightarrow \quad \det(A) = ad - bc.$$

The eigenvalue of the Jacobian matrix with the largest magnitude indicates stability according to the following theorem.

Theorem 2 *Let $J(x, y)$ be the Jacobian matrix corresponding to the two-dimensional system given in Eqs. 3 and 4 and suppose (x^*, y^*) is a fixed point. Define ρ as*

$$\rho = \max\{|\lambda_1|, |\lambda_2|\},$$

where $\lambda_{1,2}$ are the eigenvalues of $J(x^*, y^*)$. If $\rho < 1$, then (x^*, y^*) is linearly stable (attracting). If $\rho > 1$, then (x^*, y^*) is linearly unstable (repelling). If $\rho = 1$, then (x^*, y^*) is neutral or indifferent.

We note that in Theorem 2 eigenvalues of any real-valued matrix may be complex, i.e., the parabola in Eq. 8 has no x-intercepts. If an eigenvalue is complex, then the magnitude of that complex eigenvalue is the Euclidean length in the complex plane. For example, if $\lambda = \alpha + i\beta$, then

$$|\lambda| = |\alpha + i\beta| = \sqrt{\alpha^2 + \beta^2}. \tag{9}$$

If the eigenvalues are real-valued, then the magnitude is simply the absolute value.

Calculating the eigenvalues of the Jacobian matrix is the general way to determine the stability of a two-dimensional system. In fact, this method works for an n-dimensional discrete dynamical system. Therefore, if the reader is working with a three-dimensional system, finding the eigenvalues is typically the way to go. However, for two-dimensional systems, there is a shortcut we can use that bypasses the need to compute eigenvalues. The following theorem details the Jury conditions, which can be used to determine stability of a fixed point for a two-dimensional system.

Theorem 3 (Jury Conditions) *Suppose (x^*, y^*) is a fixed point of the general system given by*

$$x_{t+1} = f(x_t, y_t) \tag{10}$$

$$y_{t+1} = g(x_t, y_t). \tag{11}$$

Let $J^* = J(x^*, y^*)$, then the fixed point (x^*, y^*) is linearly stable if all three of the following conditions hold:

$$1 - tr\big(J^*\big) + det\big(J^*\big) > 0 \tag{12}$$

$$1 + tr\big(J^*\big) + det\big(J^*\big) > 0 \tag{13}$$

$$1 - det\big(J^*\big) > 0. \tag{14}$$

In each stability theorem above, we mention whether or not a fixed point is *linearly* stable. What does this mean? In any undergraduate course, one typically studies the idea of *linearization*. This is the approximation to a curve $y = f(x)$ at a point, say $x = x^*$. That is, near the point x^*, any curve looks like a line. The farther we zoom in on the curve near x^*, the more the curve looks linear. This linear approximation to the curve is the tangent line at the point x^*, whose slope is given by $f'(x^*)$. The slope tells us how the curve is changing with respect to x near the point x^*. This idea can be applied to any discrete dynamical system of the form $x_{t+1} = f(x_t)$, where we approximate the trajectory of any initial condition starting near a fixed point by a linearization of the system. In short, the magnitude of $|f'(x^*)|$ measures the degree at which this initial condition is moving toward or away from the fixed point in the linear case. This is what is meant by linear stability.

The neutral fixed point is considered a borderline case. Analytically, these points occur when the maximum eigenvalue is one, $\rho = 1$, or at least one of the Jury conditions is an equality. In fact, if this is the case, the point could still be attracting or repelling, the truth is that more information is needed about the system in order to determine stability. These points are interesting because they may indicate where (in parameter space) the system changes from stable to unstable. These ideas are developed more solidly in the following section, but please refer to [7, 21, 35], or [10] for more information regarding discrete dynamical systems.

A 2-D Example To demonstrate stability of a two-dimensional model, we consider the following example system:

$$x_{t+1} = 5 + \frac{1}{4}(x_t - y_t) \tag{15}$$

$$y_{t+1} = 2 + \frac{1}{4}x_t(1 + y_t). \tag{16}$$

We first solve for the fixed points, which means solving the following system of equations simultaneously for all ordered pairs (x^*, y^*):

$$x^* = 5 + \frac{1}{4}(x^* - y^*) \tag{17}$$

$$y^* = 2 + \frac{1}{4}x^*(1 + y^*). \tag{18}$$

Isolating x^* in the first equation, and substituting it into the second equation yields a quadratic equation in y^*, which has two solutions:

$$(y^*)^2 - 7y^* - 44 = 0 \qquad \Rightarrow \qquad y_1^* = 11, \quad y_2^* = -4. \tag{19}$$

Substituting each solution for y^* into Eq. 17 gives the following pair of fixed points:

$$(x_1^*, y_1^*) = (3, 11) \tag{20}$$

$$(x_2^*, y_2^*) = (8, -4). \tag{21}$$

Next, we seek the stability of each fixed point. Defining the right-hand side of Eqs. 15 and 16 as $f(x, y)$ and $g(x, y)$, respectively, we construct the Jacobian matrix:

$$J(x, y) = \begin{bmatrix} \frac{\partial f}{\partial x} & \frac{\partial f}{\partial y} \\ \frac{\partial g}{\partial x} & \frac{\partial g}{\partial y} \end{bmatrix} = \begin{bmatrix} \frac{1}{4} & -\frac{1}{4} \\ \frac{1}{4}(1 + y) & \frac{1}{4}x \end{bmatrix}. \tag{22}$$

Now, we evaluate the Jacobian at each fixed point, yielding two matrices:

$$J_1^* := J(x_1^*, y_1^*) = J(3, 11) = \begin{bmatrix} \frac{1}{4} & -\frac{1}{4} \\ 3 & \frac{3}{4} \end{bmatrix} \tag{23}$$

$$J_2^* := J(x_2^*, y_2^*) = J(8, -4) = \begin{bmatrix} \frac{1}{4} & -\frac{1}{4} \\ -\frac{3}{4} & 2 \end{bmatrix}. \tag{24}$$

We'll study the stability of (x_1^*, y_1^*) using the eigenvalue method. The trace and determinant of J_1^* are $\mathrm{tr}(J_1^*) = 1$ and $\det(J_1^*) = 15/16$. The eigenvalues satisfy the following equation, and can be found using the quadratic formula:

$$\lambda^2 - \lambda + \frac{15}{16} = 0 \quad \Rightarrow \quad \lambda_{1,2} = \frac{1}{2} \pm i \frac{\sqrt{11}}{4}. \tag{25}$$

Since the eigenvalues are a complex conjugate pair, the magnitudes are equal. We find the magnitude using the Euclidean norm:

$$|\lambda_{1,2}| = \sqrt{\left(\frac{1}{2}\right)^2 + \left(\pm\frac{\sqrt{11}}{4}\right)^2} = \sqrt{\frac{1}{4} + \frac{11}{16}} = \frac{\sqrt{15}}{4} < 1. \tag{26}$$

Therefore, $\rho = \max\{|\lambda_1|, |\lambda_2|\} = \sqrt{15}/4 < 1$ and the fixed point $(x_1^*, y_1^*) = (3, 11)$ is linearly stable. For the second fixed point, we'll use the Jury conditions. The trace and determinant of J_2^* are $\mathrm{tr}(J_2^*) = 9/4$ and $\det(J_2^*) = 5/16$. Considering the three conditions of Theorem 3, we find

$$1 - \frac{9}{4} + \frac{5}{16} = -\frac{15}{16} \not> 0 \tag{27}$$

$$1 + \frac{9}{4} + \frac{5}{16} = \frac{57}{16} > 0 \tag{28}$$

$$1 - \frac{5}{16} = \frac{11}{16} > 0, \tag{29}$$

which indicates that (x_2^*, y_2^*) is not linearly stable because the first condition of Theorem 3 is not satisfied. A trajectory of the system is shown in Fig. 3.

Exercise 4 Determine the stability of the fixed points of the following two-dimensional model:

$$x_{t+1} = \frac{1}{4} y_t (1 + x_t) \tag{30}$$

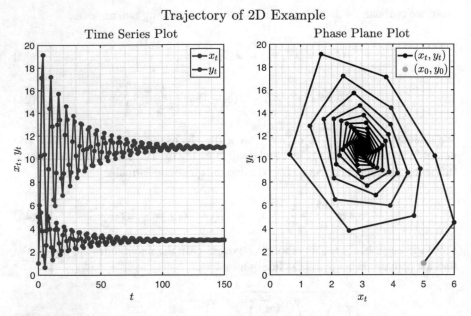

Fig. 3 A trajectory of the two-dimensional model that shows the initial condition $(x_0, y_0) = (5, 1)$ settling to the stable fixed point $(x_1^*, y_1^*) = (3, 11)$ as $t \to \infty$. See the github repository to view the MATLAB code used to produce this figure [11]

$$y_{t+1} = 3 + \frac{1}{4}(y_t - x_t). \tag{31}$$

For three-dimensional systems or higher, the eigenvalues of $J(x^*, y^*)$ are found using the cofactor expansion method. This method can be found in any introductory linear algebra text (we suggest [39]). Also, one can always find the eigenvalues of a given matrix numerically using software. For larger systems, the method of finding the magnitude of the largest eigenvalue is the method of choice for determining stability. For three-dimensional systems, stability criterion similar to that of the Jury conditions does exist. We include it in the Appendix. For all MATLAB code used to generate figures from this section, please see the chapter github repository. We refer the interested reader to [38] for information on multivariable functions and the Jacobian matrix. For more information on introductory dynamical systems and linear stability analysis, we refer the reader to [7, 21, 35], or [10]. For general linear algebra concepts, please see [39].

2 Discrete Modeling

In this section, we explore the basic discrete models of the host-parasitoid interaction. We start with a simple model of host growth with no-parasitoid influence. We then introduce one of the earliest models for the host-parasitoid system and show the general procedure for linear stability analysis. We alter the model to include a host refuge mechanism that creates stability in the original unstable system. We show how each model is built and how each model is analyzed. In the end, the reader will have a thorough understanding of the mathematical analysis involved in the modeling process. Our goal is to generalize any parametric conditions for which the coexistence (nontrivial) fixed point is linearly stable. In this way, we can effectively communicate specific model parameters that influence the coexistence of both the host and parasitoid populations.

2.1 Host-Only System

We first consider a model that contains no-parasitoid interference. In this case, only the hosts reproduce from year-to-year. We define H_t as the population of the host species in year t. We let the parameter $R > 0$ denote the number viable eggs per adult host, which implies that $R H_t$ host larvae will survive to the next year. A schematic shown in Fig. 4 shows the basic life-cycle of a typical host species.

Focusing on the discrete time-event that signifies a yearly egg production, we model the host population using a simple linear discrete model of the form

$$H_{t+1} = R H_t, \tag{32}$$

where the initial population of hosts, H_0, is known. Here, the model is linear because the right-hand side of the equation is a linear function in the previous year's host population, H_t. The host population will decay to zero if $R < 1$ and will grow without bound if $R > 1$. We are typically interested in the case when $R > 1$ because we wish to determine a form of natural biological control to help limit this growth. If $R < 1$, the host species will eventually die without the addition of a parasitic

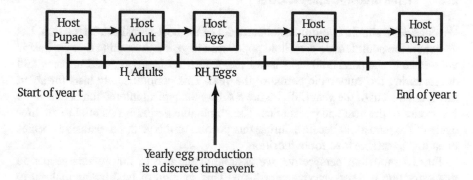

Fig. 4 A simple life-cycle schematic that shows the typical developmental stages of a host species

species. Thus, we study the implications of parasitoid suppression by introducing a female parasitoid population that limits the growth of the host species.

Exercise 5 Solve the system given in Eq. 32 for H_t by finding a way to write H_t as an expression involving only R (the host reproduction rate) and H_0 (the initial population of hosts). Evaluate $\lim_{t \to \infty} H_t$ for $R > 1$ and for $R < 1$. What other situations, biological or otherwise, may follow this same pattern?

Exercise 6 Show that the only fixed point of the model $H_{t+1} = RH_t$ is $H^* = 0$, provided $R \neq 1$. Show, using Theorem 1, that this fixed point is linearly stable if and only if $R < 1$.

Exercise 7 For $R > 1$, we see that the host-only model is unbounded, which is unrealistic as the growth of the host population must be limited in some way. Ricker's model is a one-dimensional discrete system that models the number of fish that are present year after year in a fishery [34]. Adapting this model to the host population gives

$$H_{t+1} = H_t e^{r\left(1 - \frac{H_t}{k}\right)}, \tag{33}$$

where r is interpreted as the intrinsic growth rate of the population and k is the carrying capacity. This model bounds the growth of the hosts. Can you see how? Perform a linear stability analysis on the fixed points of this model. Describe what happens as r increases.

2.2   Nicholson–Bailey Model

We consider a simple two-dimensional model of the host-parasitoid interaction. Let P_t denote the adult female parasitoid population in year t. Assuming that the host is vulnerable to the parasitoid at its larval stage, RH_t hosts are exposed to parasitoid attack during the vulnerable period of the year. This period usually lasts for about three months out of the year [20]. Figure 5 shows the typical interaction between the life-cycles of the host and parasitoids. The vulnerable period is essential to the life-cycle of the parasitoid because during this portion of the year, the parasitic species must find a suitable host for oviposition.

From a modeling perspective, we assume at the end of the vulnerable period that a fraction of hosts survive parasitism. This fraction of host larvae make it to the next developmental stage and ultimately to adulthood, where they produce R viable eggs. Denote this fraction of hosts surviving as $f(H_t, P_t)$, where we have reserved dependence on both the host and parasitoid populations. This function f is

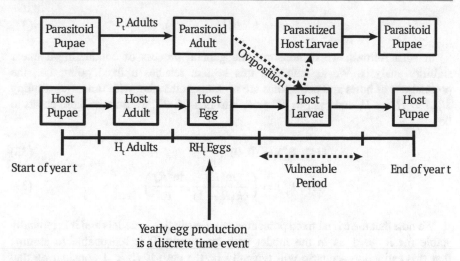

Fig. 5 A schematic to show the parallel developmental stages that make up the life-cycles of the host and parasitoid species. The transition from parasitoid adult to pupae is completely dependent on oviposition, the parasitic process of injecting eggs into the viable host larvae

assumed to be continuously differentiable whose range is [0, 1]. In some instances, this function is considered to be the *escape response* [27, 37] and it will be referred to as such from this point on. Depending on the structure of f, various nonlinear models can be formulated and each model has interesting analytical results and biological implications. Under these assumptions, we create the following two-dimensional model:

$$H_{t+1} = RH_t f(H_t, P_t) \tag{34}$$

$$P_{t+1} = kRH_t\big[1 - f(H_t, P_t)\big], \tag{35}$$

where initial populations are known. The parameter k represents the average number of female parasitoids laid on a single host that emerge and successfully become reproducing adults. Interestingly, this value of k could range anywhere from 1 to 3000, depending on the species [31, 40, 41]. Further, we can see that the number $1 - f$ is the fraction of host larvae that succumb to parasitism.

The simplest model constructed in 1935 by Nicholson and Bailey [29] assumes the escape response is a decreasing exponential function given by $f = \exp(-cP_t)$, where $c > 0$ represents the rate at which parasitoids attack the host. Indeed, for larger values of c, the fraction of host surviving decays more quickly with the density of parasitoids, P_t, indicating that the parasitoids are more efficient at searching, finding, and ovipositing into the host larvae. Substituting f into our model yields the classic Nicholson–Bailey model:

$$H_{t+1} = RH_t e^{-cP_t} \tag{36}$$

$$P_{t+1} = kRH_t\left(1 - e^{-cP_t}\right).$$ (37)

In what follows, we demonstrate the general process of conducting a linear stability analysis. We assume that this system reaches a fixed point, i.e., the populations of hosts and parasitoids are no longer changing with time. Substituting $H_{t+1} = H^* = H_t$ and $P_{t+1} = P^* = P_t$ into Eqs. 36–37, we find the fixed points to be

$$(H_0^*, P_0^*) = (0, 0)$$ (38)

$$(H_1^*, P_1^*) = \left(\frac{\ln(R)}{ck(R-1)}, \frac{\ln(R)}{c}\right).$$ (39)

We note that the trivial fixed point always exists in these models and it is typically stable for $R < 1$ as in the model with hosts-only. It is reasonable to assume that this extinction scenario will typically be the case if $R < 1$ considering that the hosts are not producing enough eggs to replenish themselves. Consequently, the parasitoid populations will go extinct because no hosts remain for oviposition. Our mathematical analysis will justify this reasoning and provide interesting results for the coexistence fixed point. Defining the right-hand side of Eqs. 36 and 37 as $F(H, P)$ and $G(H, P)$, respectively, we compute the partial derivatives in H and P (where we have suppressed the subscript t for convenience):

$$J(H, P) = \begin{bmatrix} \dfrac{\partial F}{\partial H} & \dfrac{\partial F}{\partial P} \\ \dfrac{\partial G}{\partial H} & \dfrac{\partial G}{\partial P} \end{bmatrix} = \begin{bmatrix} Re^{-cP} & -cRHe^{-cP} \\ kR(1 - e^{-cP}) & ckRHe^{-cP} \end{bmatrix}.$$ (40)

Next, we substitute the two fixed points into the Jacobian matrix to obtain the following matrices:

$$J_0^* := J(H_0^*, P_0^*) = \begin{bmatrix} R & 0 \\ 0 & 0 \end{bmatrix}$$ (41)

$$J_1^* := J(H_1^*, P_1^*) = \begin{bmatrix} 1 & -\frac{\ln(R)}{k(R-1)} \\ k(R - 1) & \frac{\ln(R)}{R-1} \end{bmatrix}.$$ (42)

By inspection, the eigenvalues of J_0^* are $\lambda_1 = R$ and $\lambda_2 = 0$. By Theorem 2, we have $\rho = \max\{|\lambda_1|, |\lambda_2|\} = |R|$, which indicates that the trivial solution is attracting for $R < 1$. Thus, for $R < 1$, extinction of both species occurs. For this reason, we typically do not consider the case when $R < 1$. A more interesting question is for what values of R or k is the coexistence fixed point, (H_1^*, P_1^*), stable? Using Theorem 3 (the Jury conditions), we arrive at the following three inequalities that *all* must be satisfied to ensure the stability of (H_1^*, P_1^*):

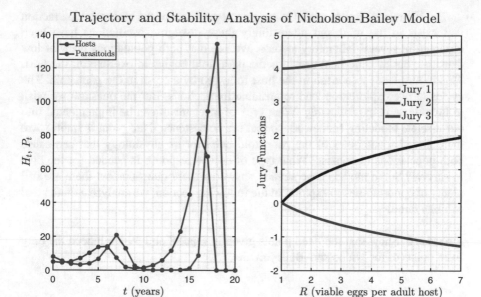

Fig. 6 (Left) A typical trajectory of the Nicholson–Bailey model that exhibits diverging oscillations. Parameters: $R = 2$, $c = 0.1$, $k = 1$, $H_0 = 5$, and $P_0 = 8$. (Right) A graphical depiction of the left-hand side of the three Jury conditions for the coexistence fixed point of the Nicholson–Bailey model. The first two conditions are always satisfied for $R > 1$, but the third condition is never satisfied. See the github repository to view the MATLAB code used to produce this figure [11]

$$\ln(R) > 0 \tag{43}$$

$$2 + \ln(R)\left(\frac{R+1}{R-1}\right) > 0 \tag{44}$$

$$1 - \ln(R)\left(\frac{R}{R-1}\right) > 0. \tag{45}$$

We note that stability is only dependent on R—the parameters c and k have no influence on whether or not the coexistence fixed point is attracting. Further, we can easily verify that the first two conditions are satisfied for $R > 1$. The last condition is *not* satisfied for $R > 1$. We note that the left-hand side of Inequalities 43–45 are all functions of R. To demonstrate that the first two conditions are satisfied but the last condition is not, we graph these *Jury functions* in Fig. 6. We can see that the first two functions are positive, but the third condition is negative for $R > 1$, suggesting that linear stability does not occur for the coexistence fixed point. Therefore, the fixed point (H_1^*, P_1^*) is linearly unstable! In other words, the Nicholson–Bailey model does not provide a condition for coexistence of the two species. Figure 6 also shows a typical trajectory of the Nicholson–Bailey model, which exhibits diverging oscillations as $t \to \infty$.

The Nicholson–Bailey model is not a good predictor of the biological interaction that exists in the wild, but interestingly these diverging oscillations have been observed in several laboratory studies. We see that both populations grow at low densities, but at large host densities the parasitoid begins to overexploit the host. This in turn leads to a crash in the host followed by a crash in the parasitoid. The cycles of overexploitation lead to unstable interactions. But the question we wish to investigate is the following: What type of mechanism can be implemented into our discrete modeling framework so that the coexistence fixed point is stable? Are there any characteristics of the parasitoid behavior or physiology that can cause both populations to coexist? What type of escape response is needed for the host to survive? Some models have already answered these questions but there are still plenty of opportunities to explore. In the following section, we analyze a model that exhibits stability.

Exercise 8 Show that the fixed points given in Eqs. 38 and 39 are indeed the only fixed points of the Nicholson–Bailey model.

Exercise 9 Derive the Jacobian matrices evaluated at each fixed point and verify that Eqs. 41 and 42 are correct. (Hint: Use $\exp(-cP^*) = 1/R$ to your advantage.)

Exercise 10 Using Theorem 3 and the Jacobian matrix evaluated at the coexistence fixed point, J_1^*, verify the three Jury conditions stated above and show graphically for all $R > 1$ that the first two conditions are satisfied but the third condition is not.

Exercise 11 We see that the Nicholson–Bailey model yields unbounded oscillations in both populations. One way to stabilize this model is to limit the growth of hosts using Ricker's model, while assuming a fraction of the hosts are parasitized by the parasitoid population. This yields the following two-dimensional system:

$$H_{t+1} = H_t e^{r\left(1-\frac{H_t}{k}\right)} e^{-cP_t} \tag{46}$$

$$P_{t+1} = H_t \left(1 - e^{-cP_t}\right), \tag{47}$$

where r and k are interpreted as in Exercise 7. This model was studied extensively by Beddington et. al. [2]. Perform a linear stability analysis on the fixed points of this model. For what values of r is the coexistence fixed point stable?

2.3   Host Refuge

One such mechanism that can stabilize the Nicholson–Bailey model is the idea of a host refuge [13, 22]. Here, we assume that a proportion, α, of the host larvae at the start of the vulnerable period manages to escape parasitism year after year. Perhaps there is a cohort of host larvae that are consistently shrouded from parasitoid attack. In other words, the parasitoid is much less effective at locating host within a certain patch or there is a proportion of hosts that evade parasitism every year. What are the conditions on this parameter α that will stabilize the system, i.e., what proportion of the available host larvae must consistently evade parasitism in order for both populations to coexist? Answering this question is the goal of our analysis—to explicitly describe the region of parameter space for which our coexistence fixed point is attracting. In this way, we seek a *stability region* in the two parameters α and R that define stability.

We let the escape response be $f = \alpha + (1-\alpha)e^{-cP_t}$, then our model becomes:

$$H_{t+1} = RH_t \left[\alpha + (1-\alpha)e^{-cP_t} \right] \tag{48}$$

$$P_{t+1} = k(1-\alpha)RH_t \left(1 - e^{-cP_t}\right). \tag{49}$$

This form of the escape response describes a host refuge because the proportion α is not parasitized, but the proportion $1 - \alpha$ is parasitized. Assuming the initial populations are given, we perform a similar stability analysis on this model. The coexistence fixed point is given by

$$(H_1^*, P_1^*) = \left(\frac{1}{ck(R-1)} \ln \left(\frac{(1-\alpha)R}{1-\alpha R} \right), \frac{1}{c} \ln \left(\frac{(1-\alpha)R}{1-\alpha R} \right) \right). \tag{50}$$

We note that this fixed point is only valid for $\alpha < 1/R$ (otherwise the fixed point is undefined). Defining the right-hand side of Eqs. 48 and 49 as $F(H, P)$ and $G(H, P)$, respectively, we compute the partial derivatives in H and P and evaluate the Jacobian matrix at the fixed points above to yield

$$J_1^* := J(H_1^*, P_1^*) = \begin{bmatrix} 1 & \left(\frac{\alpha R-1}{R-1}\right) \ln \left(\frac{(1-\alpha)R}{1-\alpha R}\right) \\ R-1 & \left(\frac{1-\alpha R}{R-1}\right) \ln \left(\frac{(1-\alpha)R}{1-\alpha R}\right) \end{bmatrix}. \tag{51}$$

We note that the Jacobian does not depend on the parasitism rate c or k, only the number of viable eggs per adult hosts, R and the proportion of initial host larvae in refuge at the start of the vulnerable period, α. Using the Jury conditions of Theorem 3, we find that linearly stable equilibria occur when

$$\alpha^* < \alpha < \frac{1}{R}, \tag{52}$$

where α^* is the solution to the following equation:

$$\frac{(1 - \alpha R)R}{R - 1} \ln\left(\frac{(1 - \alpha)R}{1 - \alpha R}\right) = 1. \qquad (53)$$

Notice that the equation above defines an implicit curve in (R, α) space. Therefore, for any given R value, the corresponding α value that satisfies the equation, denoted as α^*, can be found numerically. Once computed, the coexistence fixed point will be linearly stable for all values α between α^* and $1/R$ as stated in Eq. 52. Equation 53 is found by applying Theorem 3 (see Exercise 12), where we have changed the inequality of Jury condition 3 to an equality so that the boundary of the stability region in (R, α) space may be formally defined. In general, using software to plot implicit curves defined by the Jury conditions is the usual technique for determining the borders of a stability region. The Jury conditions provide three inequalities and so the intersection of the graphs of these inequalities will result in the stability region in the parameter space of interest.

We depict the stability region along with two trajectories in Fig. 7. In the left panel, we graph the function $\alpha = 1/R$ and the numerical values, α^*, that satisfy

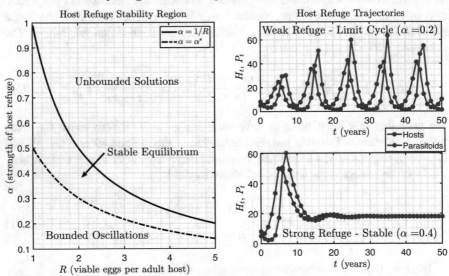

Fig. 7 (Left) A graphical depiction of the stability region in (R, α) space. If the choice of model parameters falls in between the two curves, a stable coexistence equilibrium will result. (Right) A display of host refuge model trajectories under two scenarios. (Top) With $\alpha = 0.2$, a weak host refuge exists and yields a stable limit cycle. (Bottom) A strong refuge corresponds to the α value of 0.4 and this results in a stable coexistence equilibrium. Other parameters: $R = 2$, $c = 0.1$, $k = 1$, $H_0 = 5$, and $P_0 = 8$. See the github repository to view the MATLAB code used to produce this figure [11]

Eq. 53. The two curves define the boundary of each stability region. If we choose the ordered pair $(R, \alpha) = (2, .8)$ that lies above the curve $\alpha = 1/R$, then the coexistence fixed point is undefined and our model exhibits diverging oscillations similar to the Nicholson–Bailey model. However, if we choose the ordered pair $(R, \alpha) = (2, .4)$, we land directly in between the two curves shown, which means our coexistence point is stable. We can see from the corresponding trajectory that the host and parasitoid populations approach the values H_1^* and P_1^*, respectively, as $t \to \infty$. Lastly, if we choose a pair that lies below α^*, the trajectories are neutrally stable. Therefore, the host refuge model does stabilize the system. Overall, we can report that if host reproduction is small ($R \approx 1$), then the refuge proportion must be at least half in order for stability to occur. This means that at least half of the host population has to evade parasitism year after year. Otherwise, if the host reproduction is higher ($R \gg 1$), then only a weak refuge is necessary for stability since the host population is replenishing itself at a much faster yearly rate.

Exercise 12 Using Theorem 3, show that Eq. 53 is the result of the third Jury condition.

Exercise 13 Suppose that the parasitoid population is stocked each year. Assume some constant amount is added to the population at the beginning of the vulnerable period each year. Implement this change into the Host Refuge model. How does it change the stability?

We note here that if the interested reader feels adequately versed on discrete-time models after reading this section, they can take on Research Projects 1 and 2. These projects deal with extensions to purely discrete models of host-parasitoid dynamics.

3   Continuous Modeling of the Vulnerable Period

The real-life host and parasitoid interaction is stable. This means that the two species coexist in the wild. The idea of coexistence is equivalent to our mathematical modeling having a stable fixed point of nonzero values. Mathematically, we seek relevant mechanisms that may change the qualitative dynamics from instability to stability. There are several discrete models that yield stable coexistence, like the Beddington model, for example, [20]. However, discrete models, by construction, do not recognize the decline of the host population throughout the vulnerable period— they only take into account population densities before the period begins or after the period is over. The population densities are continuously changing during this vulnerable stage of the host's life-cycle, and so we seek a model that describes specific characteristics of each population throughout this period of time. Figure 8 provides a visual representation of where we impose a continuous model. To this end, we present a general framework for modeling the dynamics of the continuous-time vulnerable period with an aim to implement the results of these continuous models into the discrete-time nature of host reproduction.

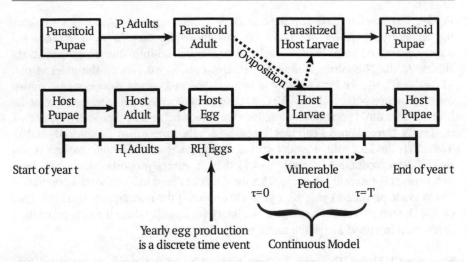

Fig. 8 The final life-cycle schematic that depicts the parallel stages of each population as well as the vulnerable period that is modeled continuously in the variable τ

3.1   General Parasitism

We implement a continuous-time model using a system of ordinary differential equations (ODEs) that describe the rate of change of three populations: unparasitized host larvae, denoted by $L(\tau, t)$; parasitized host larvae, denoted by $I(\tau, t)$; and adult female parasitoid, $P(\tau, t)$. Each function provides the current population total at time τ during the vulnerable period, in year t. The time variable t is an integer, and for each fixed t, the time variable τ ranges from $[0, T]$, where T is the length of the host vulnerable stage [29]. The biology of some species suggest that the vulnerable period is approximately one to three months [20]. This means the value of T could be anywhere from 30 to 90 days. The following reaction schematic provides a basic representation of oviposition during the vulnerable period:

$$P + L \xrightarrow{g(\cdot)} I + P. \tag{54}$$

Here, the left-hand side of this reaction depicts a parasitoid finding a potential host larvae. The function $g(\cdot)$ (units: $time^{-1} \, parasitoid^{-1}$) is the attack rate of the parasitoids, which represents the instantaneous rate at which the hosts are parasitized per parasitoid. Here, we write "$(\cdot)$" to mean that this function could potentially be dependent on the population of hosts (L), parasitoids (P), or infected hosts (I). In some cases, this rate could be assumed to be dependent on the number of hosts at the beginning of the vulnerable period. We use this notation to keep the general model as compact as possible. The right-hand side depicts the result of the interaction: a successful oviposition and an unaltered adult parasitoid female, searching for more host larvae to infect. In general, we write the continuous model as the following system of ordinary differential equations in the time variable τ:

$$\frac{dL}{d\tau} = -g(\cdot)LP \tag{55}$$

$$\frac{dI}{d\tau} = g(\cdot)LP \tag{56}$$

$$\frac{dP}{d\tau} = 0. \tag{57}$$

We note that this system assumes no host or parasitoid mortality as well as no egg limitation in the parasitoids, i.e., female parasitoids always have an egg to deposit and no transition time is wasted to accumulate enough energy to produce new eggs. We subject the system to the following initial conditions

$$L(0, t) = RH_t, \quad I(0, t) = 0, \quad P(0, t) = P_t, \tag{58}$$

where t is a fixed positive integer that represents the current year. Indeed, because the hosts lay eggs before the vulnerable stage begins and each adult host produces R viable eggs, there are RH_t available host larvae at the start of the vulnerable period. Further, at the start of any vulnerable period, no larvae have been infected and the number of female adult parasitoids is known to be P_t. Given an expression for the attack rate $g(\cdot)$, these equations are integrated from $\tau = 0$ to $\tau = T$ to produce a profile of the population totals throughout the vulnerable period. We note here that these solutions for L, I, and P give the population values *during a single year* throughout the vulnerable period. Therefore, the values given by $L(T, t)$ and $I(T, t)$ are the numbers of un-parasitized and parasitized host larvae, respectively, at the end of the vulnerable period for year t.

We now have a framework for the vulnerable period that allows for opportunity as well as creativity. The parasitism rate, defined by $g(\cdot)$, can be as simple or complicated as we make it. In short, this parasitism function may incorporate many factors such as the searching efficiency of the parasitoid as well as the time it takes to make a successful oviposition or find a suitable host. What kind of relevant characteristics can we include in this parasitism rate? Or what additional, relevant reactions can be included in the model? We explore two models that yield explicit solutions to the system in Eqs. 55–57 and we explore a model that implements an additional reaction in Sect. 3.2.

3.1.1 Constant Rate of Attack

We consider a simple model where the parasitism rate is constant, i.e., $g(\cdot) = c$. Substituting this constant into Eqs. 55–57, we obtain the system

$$\frac{dL}{d\tau} = -cLP \qquad L(0, t) = RH_t \tag{59}$$

$$\frac{dI}{d\tau} = cLP \qquad\qquad I(0,t) = 0 \tag{60}$$

$$\frac{dP}{d\tau} = 0 \qquad\qquad P(0,t) = P_t. \tag{61}$$

This is a system of nonlinear ODEs whose solutions can be found explicitly. Note that the rate of change of the parasitoid population is zero, indicating that the parasitoids remain constant throughout the vulnerable period. This fact together with the initial parasitoid population indicates that $P(\tau,t) = P_t$. Thus, no parasitoids are lost during the vulnerable period and the number of parasitoids remains at P_t throughout year t. This assumption can be relaxed in a more dynamic model (see Exercise 19). Substituting P_t into our equation for L yields the following linear ODE:

$$\frac{dL}{d\tau} = -cP_t L. \tag{62}$$

This equation is solved using separation of variables to yield $L(\tau,t) = RH_t e^{-cP_t \tau}$, where we have incorporated the initial condition to eliminate the integration constant. To solve for I, we note that $\frac{dI}{d\tau} + \frac{dL}{d\tau} = 0$, indicating that $I + L$ is constant throughout the vulnerable period. Hence, enforcing the initial conditions of I and L, we obtain $I(\tau,t) = RH_t \left(1 - e^{-cP_t \tau}\right)$. Therefore, we have the following solutions:

$$L(\tau,t) = RH_t e^{-cP_t \tau} \tag{63}$$

$$I(\tau,t) = RH_t \left(1 - e^{-cP_t \tau}\right) \tag{64}$$

$$P(\tau,t) = P_t. \tag{65}$$

When the parasitism rate is constant, we can predict the exact population densities at any time τ during the vulnerable period. Figure 9 shows two profiles of these solutions for varying parasitism rate c. We can see that as c increases, the efficiency of the parasitoids to oviposit increases. This results in a sharp increase in the number of infected host larvae during the vulnerable period. This shows that for a larger value of c, more host larvae succumb to parasitism.

Exercise 14 Verify that the solutions given in Eqs. 63–65 are indeed solutions to the constant parasitism ODE system.

3.1.2 Functional Response

In more realistic scenarios the parasitoid attack rate is not constant. The attack rate, denoted generally as $g(\cdot)$, can be modeled as a function of host density. If we let this function be an increasing function of L, then the model assumes that the parasitoids

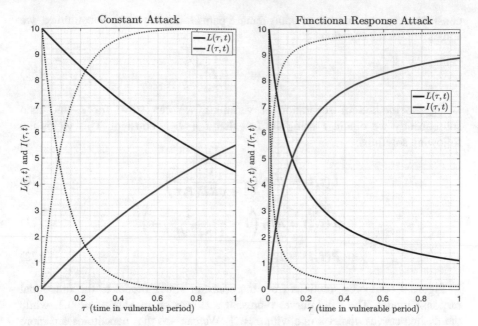

Fig. 9 A time-series plot of the population densities of $L(\tau, t)$ and $I(\tau, t)$ during the vulnerable period for the constant attack model (left) and the functional response model (right). The solid curves depict a slower attack rate, $c = 0.1$. The dashed curves depict a faster attack rate $c = 1$. Parameters: $R = 2$, $H_t = 5$, $P_t = 8$, $T = 1$. See the github repository to view the MATLAB code used to produce this figure [11]

get better at finding a host larvae if there are more host larvae available. This seems reasonable. This type of attack rate is typically called a *functional response* because the parasitoids adjust their attack based on the current host larvae population. The seasonal variation in host larvae has an impact on the parasitoid's search efficiency.

The simplest increasing function of L is a linear function of the form $g(L) = cL$, where c is the slope. We can interpret the slope as the parasitoids attack response to any change in the host density. Substituting this function into Eqs. 55–57 yields the following ODE system:

$$\frac{dL}{d\tau} = -cL^2 P \qquad L(0, t) = RH_t \tag{66}$$

$$\frac{dI}{d\tau} = cL^2 P \qquad I(0, t) = 0 \tag{67}$$

$$\frac{dP}{d\tau} = 0 \qquad P(0, t) = P_t. \tag{68}$$

This system can be solved explicitly as in the previous section, using similar tricks. Recalling that τ is the continuous-time variable describing the vulnerable period, $P(\tau, t)$ is not changing during this period, which suggests $P(\tau, t)$ remains constant at P_t, the number of adult female parasitoids in year t. Substituting, we solve for L in the equation

$$\frac{dL}{d\tau} = -cP_tL^2.$$ (69)

Using the method of separation of variables, we can solve for L and then subsequently we obtain I using the fact that $I + L$ is constant. This yields the following solutions:

$$L(\tau, t) = RH_t \left(\frac{1}{1 + cRH_t P_t \tau} \right)$$ (70)

$$I(\tau, t) = RH_t \left(1 - \frac{1}{1 + cRH_t P_t \tau} \right)$$ (71)

$$P(\tau, t) = P_t.$$ (72)

Figure 9 compares solutions from the constant attack model to the functional response model. The solid curves represent a slower attack rate, $c = 0.1$, while the dashed curves depict a rate with $c = 1$. We can see that parasitoids are more effective at parasitizing the host larvae under the functional response model.

Exercise 15 What other types of functional responses could we implement into the reaction and still be able to solve the ODE system explicitly? Even if the equations cannot be solved explicitly, solve any new model numerically and output your solutions on a graph to compare whether the parasitoid is more or less efficient at attacking the host during the vulnerable period.

Exercise 16 Use the method of separation of variables to arrive at the solutions given by Eqs. 70–71.

Exercise 17 Assume $g(L) = cL^m$, where c is constant and m is a positive integer with $m \geq 1$. Solve the system of ODEs explicitly.

3.2   Host Mortality

The general model framework provided above lends itself to biologically relevant additions. In short, we do not have to restrict ourselves to parasitism only— there may be other host or parasitoid characteristics that can be incorporated. One mechanism that is important to consider is host mortality due to factors other than parasitism. As stated in [17], a variety of parasitoid species may oviposit in the same type of host. Also, other natural factors may lead to the destruction of the host population such as severe weather or other predators. Therefore, in this model, we incorporate a host death at a rate $g_d(\cdot)$ that may or may not be density dependent. We start by considering the following reaction schematic:

$$P + L \xrightarrow{g(\cdot)} I + P \tag{73}$$

$$L \xrightarrow{g_d(\cdot)} \text{Death.} \tag{74}$$

From the schematic, we see that the total host larvae population, uninfected and infected, is no longer conserved. In particular, conservation of hosts is lost because hosts are being depleted by something else that is not in the system rather than being transferred from one state (uninfected) to another (infected). This means that solving for I will be an additional task! Reading the schematic, we arrive at the following general form of the ODE system:

$$\frac{dL}{d\tau} = -g(\cdot)LP - g_d(\cdot)L \tag{75}$$

$$\frac{dI}{d\tau} = g(\cdot)LP \tag{76}$$

$$\frac{dP}{d\tau} = 0 \tag{77}$$

subject to identical initial conditions

$$L(0, t) = RH_t, \quad I(0, t) = 0, \quad P(0, t) = P_t. \tag{78}$$

Letting the parasitism rate be constant, $g = c$, and the host mortality rate be dependent on the host density, $g_d(L) = c_d L$, we can solve the ODE system to yield

$$L(\tau, t) = RH_t \left[\frac{1}{e^{cP_t\tau} + c_d RH_t \left(\frac{e^{cP_t\tau} - 1}{cP_t} \right)} \right] \tag{79}$$

$$I(\tau, t) = \frac{cP_t}{c_d} \ln \left[1 + c_d RH_t \left(\frac{1 - e^{-cP_t\tau}}{cP_t} \right) \right] \tag{80}$$

$$P(\tau, t) = P_t. \tag{81}$$

The nature of these solutions breaks the general trend of the previous two models. This is a result of the fact that there are much less host larvae available throughout the vulnerable period due to additional host death. Figure 10 shows the time-series plot of the uninfected and infected host larvae. We can see that the total population is not conserved—the total host population is at a steady decline. It is clear that over

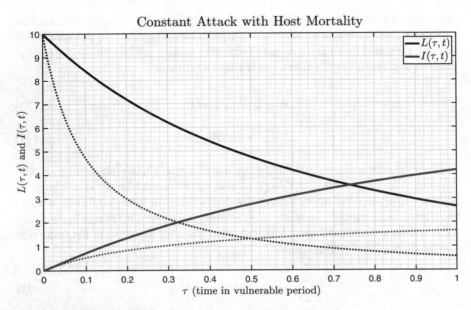

Fig. 10 A time-series plot of the population densities of $L(\tau, t)$ and $I(\tau, t)$ during the vulnerable period for the constant attack model with density dependent host mortality. Both sets of curves have a parasitism rate of $c = 0.1$. The solid curves depict a slower host mortality rate, $c_d = 0.1$. The dashed curves depict a faster host mortality rate $c_d = 1$. Parameters: $R = 2$, $H_t = 5$, $P_t = 8$, $T = 1$. See the github repository to view the MATLAB code used to produce this figure [11]. Two programs are included: one that plots the explicit solutions and another that solves the system numerically

a single vulnerable period, the additional host mortality has a negative effect on both species. However, we could speculate that the host mortality may indeed help the host population in the long run considering this short term lost due to host mortality allows less parasitism to occur and so the initial parasitoid population in the next cycle is suppressed.

Exercise 18 Verify that the solutions given in Eqs. 79–81 are indeed solutions to the constant ODE system.

Exercise 19 Create a general model that has a parasitoid mortality, whose rate is given by the function $g_p(\cdot)$. Develop the ODE system corresponding to this scenario. Suppose this rate is constant, $g_p = c_p$. Solve the system explicitly and plot the solutions. Compare the solutions to the model in Eqs. 63–65.

3.3   Host-Feeding

In the previous model, we included an additional reaction to simulate density dependent mortality. Until now, we have only defined three species that interact over the vulnerable period: uninfected host larvae, infected host larvae, and female parasitoids. Must we limit ourselves to only these three species? One interesting dynamic of the parasitoid is the behavior known as host-feeding. This occurs when a parasitoid consumes a potential host larvae without ovipositing [23–25]. In fact, a parasitoid may partially destroy a host and still oviposit inside of it. Therefore, we make our model as general as possible and include several types of host-feeding behavior. To do this, we introduce additional variables.

We define four populations that evolve during the vulnerable period: $L(\tau, t)$ is the population of uninfected hosts, $I_0(\tau, t)$ is the population of infected hosts that are partially destroyed due to host-feeding, $I_1(\tau, t)$ is an intact infected host, and $P(\tau, t)$ is the female parasitoid population. We consider the following reaction kinetics that describe the uninfected and infected host-feeding dynamic:

$$P + L \xrightarrow{g(\cdot)} P + I_1 \tag{82}$$

$$P + L \xrightarrow{g_h(\cdot)} P \tag{83}$$

$$P + L \xrightarrow{g_o(\cdot)} P + I_0 \tag{84}$$

$$P + I_1 \xrightarrow{g_i(\cdot)} P + I_0. \tag{85}$$

Here, reaction in Eq. 83 describes un-parasitized host-feeding where a healthy host is completely consumed. Reaction 84 is partial host-feeding and the reaction depicted in Eq. 85 describes infected host-feeding, where the parasitoid partially consumes the infected host, but does not kill it, which yields a partially destroyed infected host. This partially destroyed host will still produce parasitoid larvae in the next year. The reaction kinetics are translated to the follow system of ODEs:

$$\frac{dL}{d\tau} = -[g(\cdot) + g_h(\cdot) + g_o(\cdot)] LP \tag{86}$$

$$\frac{dI_0}{d\tau} = g_o(\cdot)LP + g_i(\cdot)I_1P \tag{87}$$

$$\frac{dI_1}{d\tau} = g(\cdot)LP - g_i(\cdot)I_1P \tag{88}$$

$$\frac{dP}{d\tau} = 0, \tag{89}$$

subject to

$$L(0, t) = RH_t, \quad I_0(0, t) = 0, \quad I_1(0, t) = 0, \quad P(0, t) = P_t. \tag{90}$$

Now, we have a mathematical model for host-feeding. The goal is to determine what kind of rate functions will yield analytical solutions to the differential equations. We leave the constant rate function scenario up to the reader in the following exercise.

Exercise 20 In the host-feeding model, let every rate function be constant. Solve the system of ODEs and compare the solutions to the model in Eqs. 63–65 that result from constant parasitism alone. Further, plot the solutions for L, I_0, I_1, and P for $\tau \in [0, T]$.

3.4 Other Mechanisms

Many other mechanisms can be incorporated into the ODE system during the vulnerable period. In fact, this is the beauty of the continuous system—there is freedom to create any scenario that is plausible. For example, we could consider a simple egg-load capacity [8, 27] by defining two types of parasitoids: $P_0(\tau, t)$ and $P_1(\tau, t)$, where the subscript denotes the number of eggs the female parasitoid currently has for deposit. Using this definition, we can develop the following reactions:

$$P_1 + L \xrightarrow{g(\cdot)} P_0 + I \tag{91}$$

$$P_0 \xrightarrow{g_r(\cdot)} P_1. \tag{92}$$

Here, we see that a parasitoid can only parasitize a host larvae if it has an egg to deposit. Otherwise, the parasitoid is in transition between no egg and with egg. The transition period may be characterized by the need for the parasitoid to gain energy to produce another egg or simply to have the strength to search and oviposit into another host. The question, then, for this model is how does this rate g_r effect the overall dynamics of the system? Are the equations still solvable? What are the initial conditions for P_0 and P_1?

Exercise 21 In the egg delay model above, let each reaction rate be constant. Set up the differential equations and solve them numerically using the MATLAB code `Egg_Delay_ODE` in the github repository. Describe the nature of the solutions for $g_r \to 0$ and for $g_r \to \infty$.

Another interesting generalization of this model is the idea of a host patch system. Perhaps the hosts collect in patches and these locations make up a patch network [5, 9, 18]. Parasitoids are able to fly from patch to patch but hosts are immobile, residing at the same patch throughout the vulnerable period. Define $H_i(\tau, t)$, $I_i(\tau, t)$, and $P_i(\tau, t)$ as the population of hosts, infected hosts, and parasitoids at location i on a one-dimensional patch grid. Assume the parasitoids can fly from patch i to patch j at a rate m_{ij}. Should the rate m_{ij} be density dependent? What type of migration characteristics influence the dynamics the most?

We have discussed some of these characteristics above but there are many other things to consider. We refer the interested reader to [13, 32, 33, 36, 41] for a more thorough treatment of the behaviors of parasitoid species.

Exercise 22 Construct the reaction schematic that would describe two parasitoids, P and Q, parasitizing a single host larvae population, L. Then, create a reaction that would represent one parasitoid, say P, parasitizing a host that has been infected by Q. This new reaction converts the Q-infected host to a P-infected host. Set up the ODEs with constant parasitism for all. Solve them explicitly, if possible. Plot the solutions over the vulnerable period.

Exercise 23 Construct the reaction schematic that would describe a community of three parasitoid populations and two host populations where one of the parasitoids can parasitize either host (i.e., a generalist) and the other two parasitoids can only parasitize their respective host populations (i.e., specialists). Set up the ODEs with constant parasitism for all. Solve them explicitly, if possible. Plot the solutions over the vulnerable period.

4 Semi-discrete Modeling

In this section, we merge the discrete modeling technique of Sect. 2 with the continuous modeling framework of the vulnerable period of Sect. 3 to form a hybrid model. As we will see, the hybrid or semi-discrete model is actually discrete in nature but the right-hand side functions are determined completely by the results of the ODE model during the vulnerable period. Therefore, we typically use the same analytical methods of Sect. 2, but we create the model from the explicit solutions to the ODEs. In this way, the influence of any mechanisms that we introduce during the vulnerable period is realized on the yearly time-scale in the discrete model.

4.1 General Framework

As in the discrete modeling approach, we assume that the host and parasitoid density in year t is given by H_t and P_t, respectively. We also assume that the functions $L(\tau, t)$ and $I(\tau, t)$ describe the populations of uninfected and infected host larvae, respectively, during the year t vulnerable period, where $\tau \in [0, T]$. Here, the value T is a parameter that represents the total length of the vulnerable period. For our purposes, we generally take the value of T to be one to represent a single vulnerable period. The biology of some species suggest that the vulnerable period is approximately one to three months [20]. Therefore, the quantities $L(T, t)$

and $I(T, t)$, denote the total number of un-parasitized host larvae and parasitized host larvae, respectively, at the end of the vulnerable period. The discrete model predicts the amount of each population in the next year assuming a constant host reproductive rate, given by R, and an average number of parasitoid larvae, k, emerging from the host at the end of the vulnerable period. Thus, our semi-discrete model is formulated as follows:

$$H_{t+1} = L(T, t) \tag{93}$$

$$P_{t+1} = kI(T, t). \tag{94}$$

Note that the parameter R does not appear in the top equation. This is because R is already incorporated in the initial condition for the ODE system, and so $L(T, t)$ will have the proper dependence on R. Also, we do multiply the quantity $I(T, t)$ by k because more than one parasitoid can emerge from a single parasitized host [31, 40]. Notice that the right-hand side functions of the discrete model are the direct results of the ODE system. Therefore, being able to find an explicit solution to the ODE system is critical in the development of our semi-discrete approach. However, you need not have explicit solutions to explore the model numerically. To obtain analytical results about the fixed points of the system, explicit solutions are necessary.

In the sections to follow, we explore a few models that are direct consequences of the continuous methods considered in the previous section.

4.1.1 Constant Attack

Consider the solutions to the constant attack model of Sect. 3.1.1. These solutions represent the number of uninfected and infected host larvae over the span of the vulnerable period, which starts at $\tau = 0$ and ends at $\tau = T$. Thus, substituting $\tau = T$ into the explicit solutions given by Eqs. 63 and 64 yields the number of hosts that will survive into the next year, $H_{t+1} = L(T, t)$, and the number of parasitioids emerging into the next year, $P_{t+1} = kI(T, t)$. This gives the following semi-discrete yearly update:

$$H_{t+1} = RH_t e^{-cP_t T} \tag{95}$$

$$P_{t+1} = kRH_t \left(1 - e^{-cP_t T}\right). \tag{96}$$

Notice anything familiar about this system? This system is essentially identical to the Nicholson–Bailey model (Eqs. 36 and 37)! The only difference is that T appears in the exponential escape response, $f = \exp(-cP_t T)$. The stability of this model is equivalent to the Nicholson–Bailey model. The Nicholson–Bailey model is unstable for all c and R, where the parameter c appears next to P_t in the exponential function. Because this model includes the lumped coefficient cT, we can assume that T has no real effect on the stability of the system. Therefore, the constant attack semi-discrete model is unstable and yields diverging oscillations.

We see that implementing a constant attack rate in the vulnerable period (the most simple type of response) results in a semi-discrete model that is identical to the Nicholson–Bailey model. This begs the question: What type of additional mechanisms can be incorporated in the vulnerable period that will yield a stable semi-discrete model? We explore a few cases in the sections that follow.

4.1.2 Functional Response

Following the same procedure, we consider the functional response attack rate of Sect. 3.1.2. Substituting the explicit solutions in Eqs. 70 and 71 evaluated at $\tau = T$ yields the following semi-discrete update:

$$H_{t+1} = RH_t \left(\frac{1}{1 + cRH_t P_t T} \right) \tag{97}$$

$$P_{t+1} = kRH_t \left(1 - \frac{1}{1 + cRH_t P_t T} \right). \tag{98}$$

Here, we see the escape response is dependent on both H_t and P_t, given by

$$f(H_t, P_t) = \frac{1}{1 + cRH_t P_t T}. \tag{99}$$

This is a decreasing function of P_t with $f(H_t, 0) = 1$, indicating that all host larvae survive to the next year in the absence of parasitoids. Solving for the fixed points of this system, we obtain the trivial fixed point as well as a nontrivial fixed point given by

$$(H_1^*, P_1^*) = \left(\frac{1}{\sqrt{kcRT}}, \frac{\sqrt{k}(R - 1)}{\sqrt{cRT}} \right). \tag{100}$$

Defining F and G as the right-hand side functions of the system above in Eqs. 97 and 98, we compute the Jacobian matrix evaluated at (H_1^*, P_1^*) below:

$$J_1^* := J(H_1^*, P_1^*) = \begin{bmatrix} \frac{1}{R} & -\frac{1}{kR} \\ k(R - 1)\left(1 + \frac{1}{R}\right) & \frac{1}{R} \end{bmatrix}. \tag{101}$$

A quick calculation reveals the trace and determinant of the Jacobian matrix as

$$\operatorname{tr}(J_1^*) = \frac{2}{R} \qquad \det(J_1^*) = 1. \tag{102}$$

Using this information, we'll show that this fixed point is neutrally stable in two different ways. First, we can examine the eigenvalues of J_1^* by considering Theorem 2. The eigenvalues of this matrix, denoted by λ, are roots of the quadratic polynomial defined by $\lambda^2 - \text{tr}(J_1^*)\lambda + \det(J_1^*)$. Substituting the trace and determinant into this quadratic and applying the quadratic formula gives the following eigenvalues:

$$\lambda^2 - \frac{2}{R}\lambda + 1 = 0 \quad \Rightarrow \quad \lambda_{1,2} = \frac{1}{R} - \sqrt{\frac{1}{R^2} - 1}. \tag{103}$$

Since we are only considering the case when $R > 1$ (the host population needs a yearly reproduction of at least one egg per adult host otherwise the host population dies out), we see that the eigenvalues are complex with real part $1/R$ and imaginary part $\sqrt{1 - 1/R^2}$. Computing the length of these complex eigenvalues yields

$$|\lambda_{1,2}| = \sqrt{\left(\frac{1}{R}\right)^2 + \left(\sqrt{1 - \frac{1}{R^2}}\right)^2} = 1. \tag{104}$$

Therefore, both eigenvalues have unit magnitude, $\rho = 1$, and the fixed point given by Eq. 100 is neutral. Another way to show that this fixed point is neutrally stable is to investigate the three Jury conditions of Theorem 3. Using the trace and determinant directly, we see that in order for the coexistence fixed point to be stable, the following three conditions must hold:

$$1 - \text{tr}(J_1^*) + \det(J_1^*) > 0 \qquad\qquad 2 - \frac{2}{R} > 0 \tag{105}$$

$$1 + \text{tr}(J_1^*) + \det(J_1^*) > 0 \quad \Rightarrow \qquad 2 + \frac{2}{R} > 0 \tag{106}$$

$$1 - \det(J_1^*) > 0 \qquad\qquad\qquad 0 \not> 0 \tag{107}$$

The first two conditions are satisfied for all $R > 1$. The third condition is never satisfied. The determinant of the Jacobian matrix is exactly 1, which means the third Jury condition is the most restrictive condition. Since the third condition is zero, the model lies on the boundary of being stable and unstable. Therefore, we say the fixed point is neutrally stable. This means that all trajectories corresponding to the functional response model are oscillatory in nature. The host and parasitoids are trapped in a *limit cycle*, similar in nature to the weak host refuge model of Sect. 2.3. Figure 11 shows a trajectory of the functional response model. We see that the host and parasitoids experience repeated growth and decay in a cyclic behavior similar to the Nicholson–Bailey model but the amplitude stays bounded.

We conclude that a linear function response in the attack can stabilize the Nicholson–Bailey model, but not quite enough to result in a stable fixed point. Higher-order functional responses can yield stability. We ask the reader to verify this in Exercise 24.

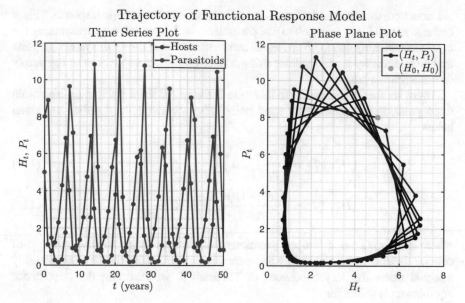

Fig. 11 A time-series and phase plane trajectory of the linear functional response model that results in a limit cycle. Parameters: $R = 2$, $c = 0.1$, $k = 1$, $T = 1$, $H_0 = 5$, and $P_0 = 8$. See the github repository to view the MATLAB code used to produce this figure [11]

Exercise 24 Construct the yearly update using the solution from Exercise 17. Then, perform an identical stability analysis using a general functional response attack rate. Is the model stable for $m = 2$? What happens to the fixed point values H^* and P^* as m increases? Are the parasitoids more or less efficient?

4.2 Host Mortality

We follow the same procedure for the host mortality model. Consider the explicit solutions in Eqs. 79 and 80 from the continuous host mortality model of Sect. 3.2. The discrete dynamical system corresponding to the host mortality characteristic is below:

$$H_{t+1} = RH_t \left[\frac{1}{e^{cP_t T} + c_d RH_t \left(\frac{e^{cP_t T} - 1}{cP_t} \right)} \right] \tag{108}$$

$$P_{t+1} = k \frac{cP_t}{c_d} \ln \left[1 + c_d RH_t \left(\frac{1 - e^{-cP_t T}}{cP_t} \right) \right]. \tag{109}$$

We note here that this model breaks the general trend of an escape response. This is because not all hosts succumb to just parasitism. In this model, we are assuming the hosts are perishing due to other means, and so the result of the ODE system suggests that host mortality is not modeled using a fraction of host survival as in the purely discrete framework.

There are three fixed points to this model: the trivial fixed point, a no-parasitoid fixed point, and a coexistence fixed point. The nontrivial fixed points are given below:

$$(H_1^*, P_1^*) = \left(\frac{R-1}{c_d RT}, 0 \right) \tag{110}$$

$$(H_2^*, P_2^*) = \left(\frac{(e^z - 1)(\ln R - z)}{c_d(R - e^z)T}, \frac{\ln R - z}{cT} \right), \tag{111}$$

where $z = c_d/kc$ is a lumped parameter that represents the relative strength of density dependent host mortality and parasitism. Performing the standard stability analysis using the Jury conditions of Theorem 3, we find that the coexistence equilibrium is stable for

$$z^* < z < \frac{\ln R}{ck}, \tag{112}$$

where z^* is the solution to the following equation

$$z + 1 = \frac{R(\ln R - z)}{R - e^z}. \tag{113}$$

The double inequality above defines a stability region in (R, z) space. This stability region partitions the plane into three zones given in Fig. 12. Suppose R is fixed at 2. Then we see that if $z = 0.23$, the model will result in bounded oscillations. If $z = 0.5$, a stable coexistence equilibrium results suggesting that the ratio of host mortality and parasitism is balanced to sustain both populations. However, if $z = 0.8$, the host mortality due to other factors is more prevalent within the system and the parasitoid population dies out. Doesn't this seem counterintuitive? If the hosts are dying more, shouldn't they be the ones to die out? But remember the parasitoids need the hosts to survive; therefore, a threshold mortality rate exists in order for the parasitoids to have enough hosts for implantation. In short, the parasitoids are competing with an outside factor for the host resource. If the lumped parameter, z, is too large, then the parasitoids are consuming the host resource at a slower rate than the other competitors.

Exercise 25 Prove that there are indeed three fixed points to the host mortality model and show that Eqs. 108 and 109 are the nontrivial fixed points. Use the Jury conditions to arrive at Eq. 113.

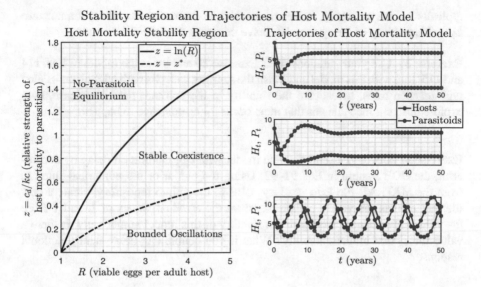

Fig. 12 (Left) Stability region of the host mortality model in (R, z) space, where z is the ratio of host mortality to parasitism. (Right) Three trajectories that show a result of each stability zone: No-parasitoid equilibrium, $z = 0.8$(top); coexistence, $z = 0.5$ (middle); and bounded oscillations, $z = 0.23$ (bottom). Other parameters: $R = 2$, $c = 0.1$, $k = 1$, $T = 1$, $H_0 = 5$, and $P_0 = 8$. See the github repository to view the MATLAB code used to produce this figure [11]

4.3   Host-Feeding

We end this section with a cliffhanger. The authors in [12] consider a host-feeding dynamic with an egg-load capacity. However, they do not consider a general host-feeding scenario where the parasitoid may feed on healthy or infected hosts. As mentioned previously there is a tendency of some parasitoid species to feed on the host sometimes before, during, or after oviposition [24]. This poses an interesting question: what happens if parasitoid feeds on an already infected host?

Using the solutions to Exercise 20, and assuming each full intact parasitized host larvae gives rise to k_1 adult parasitoids in the next generation and each partially destroyed parasitized host larvae gives rise to a reduced number of parasitoids k_0, the semi-discrete framework formulation for the yearly update of hosts and parasitoids is given by

$$H_{t+1} = L(T, t) \tag{114}$$

$$P_{t+1} = k_1 I_1(T, t) + k_0 I_0(T, t). \tag{115}$$

We note here that if $k_0 = 0$, then the parasitoids are destroying the infected host larvae to the point where the parasitoid larvae does not survive into the next year. This is analogous to the parasitoid eating their own undeveloped offspring, which is typically not the case [14]. However, we wish to determine the effect of a nonzero k_0 on the stability of the system. We leave this as an exercise for the reader.

Exercise 26 Use your solutions from Exercise 20 and build the model in Eqs. 114 and 115. Run with this model and finish the stability analysis with Nicholson–Bailey type reaction rates. Investigate the stability in the constant parasitism case, then explore different mechanisms that must occur for coexistence to happen.

Exercise 27 Develop a semi-discrete model with egg maturation delay by considering the ODE system in Eqs. 91–92. Using the output of the numerical solution from the MATLAB file `Egg_Delay_ODE` in the github repository, create a new file that models the yearly update based on the terminal values of L and I from the vulnerable period. Investigate the stability of this model numerically. Is there a value for g_r that induces stability? What if you couple egg delay with functional response?

5   Suggested Research Projects

There are a number of interesting directions we can take this semi-discrete modeling framework. Now that we have a solid foundation on the host-parasitoid models, we hope to explore a few other interesting ideas. We propose four projects, two of which are purely discrete in nature that can be formulated based the Nicholson–Bailey, functional response, or host refuge models described above and the other two are semi-discrete in nature, where a formulation of an ODE system is required. We present each project below and include a bulleted list of the mathematical methods needed for success.

Discrete Modeling Projects

Research Project 1 Develop a general model for a two-parasitoid (P_t and Q_t), one-host (H_t) system with a general escape response from each parasitoid species, say f and g, into the next year. In some parasitoid species, the first few parasitoid larvae that develop inside the host larvae have special capabilities. They are unable to reproduce but have large mandibles. They develop more quickly and play the role of a "defender" while the other, normal larvae continue to develop. That is, the first developers will attack any other parasitoid egg, from a different species, that is within the same host [15,41]. Once you have a model of two-parasitoid, one-host formulated, create a variant that implements this defender capability by one of the parasitoid species.

Research Project 2 Develop a general model for a host-parasitoid system that includes a hyperparasitoid population that oviposits into the infected host larvae during a later, disjoint vulnerable period. Assume an escape response for hosts, denoted by f, and an escape response for the infected host larvae, denoted by g, into the next year. Under Nicholson–Bailey dynamics, is the model stable? If not, what can you introduce into the model to stabilize it? Experiment with different escape response functions.

In particular, for Research Projects 1 and 2, you will have to:

- Formulate a three-dimensional discrete system that incorporates the behavior of a second parasitoid population, either another parasitoid or a hyperparasitoid.
- Analyze a three-dimensional system using either the eigenvalue method or the Jury conditions for three-dimensional systems in the Appendix.
- First determine stability regions for the simplest model—constant attack rates. Then, explore different types of escape responses that may include functional response or host refuge.
- Develop well-annotated figures and graphs that help to clarify and justify your analytical results.

Semi-Discrete Modeling Projects

Research Project 3 Develop a community model with two specialist parasitoids and one generalist parasitoid. That is, formulate a model with three parasitoid species and two host species, where one parasitoid species can parasitize both hosts, but the other parasitoids can only parasitize one of the hosts. Explore what mechanisms yield stability in this scenario.

Research Project 4 Develop a general model of a network of host-parasitoid interactions where the uninfected host, infected host, and parasitoid populations at patch i are denoted by $H_i(\tau, t)$, $I_i(\tau, t)$, and $P_i(\tau, t)$, respectively. For simplicity, assume that only basic parasitism occurs as each site with constant attack rate. Suppose that parasitoids at patch i may leave to search in an adjacent patch j and they can move between adjacent patches at a rate m_{ij}. Solve the resulting system of equations, if possible, and determine the effects of the patch network on the dynamics of the system.

In particular, for Research Projects 3 and 4, you will have to:

- Formulate an n-dimensional discrete system that is the result of a specific ODE system that reflects either the competition or general patch network migration.
- Analyze each model using either the eigenvalue method or the Jury conditions that we used for the models above.
- Determine the stability regions for the simplest model—constant attack rates. Then, explore different types of escape responses that may include functional response or host refuge.
- Develop well-annotated figures and graphs that help to clarify and justify your analytical results.

For Research Project 3, you will need to come up with a five dimensional model, which means maximum eigenvalue approach to stability analysis will be used. For Research Project 4, you will need to develop an n-dimensional ODE system that includes parasitoid migration among an n-patch network. The general solution to this ODE system will be difficult to explicitly state, but it is possible. Once you have the solutions, you can formulate the two-dimensional discrete system and perform the stability analysis.

Appendix: Stability Theorem for 3-D Systems

The theorem below describes the stability criterion for three-dimensional discrete systems.

Theorem 4 *Let J^* be the Jacobian matrix for a three-component discrete system at a fixed point. Let J_k^* be the 2×2 matrix obtained from J^* by deleting row k and column k. Define c_1, c_2, and c_3 by*

$$c_1 = -tr(J^*)$$

$$c_2 = \sum_{i=1}^{3} det(J_k^*)$$

$$c_3 = -det(J^*).$$

Then the fixed point of the original system is asymptotically stable if

$$1 + c_1 + c_2 + c_3 > 0 \tag{116}$$

$$1 - c_1 + c_2 - c_3 > 0 \tag{117}$$

$$1 - c_3^2 - |c_2 - c_1 c_3| > 0. \tag{118}$$

References

1. van Alphen J. J. M., Bernstein, C., Driessen, G.: Information acquisition and time allocation in insect parasitoids. Trends in Ecology and Evolution **18**(2), 81–87 (2003)
2. Beddington, J.R., Free, C.A., Lawton, J.H.: Dynamic complexity in predator-prey models framed in difference equations. Nature **255**(5503), 58–60 (1975)
3. Bonsall, M.B., Hassell, M.P.: Parasitoid-mediated effects: apparent competition and the persistence of host-parasitoid assemblages. Popul. Ecol. **41**(1), 59–68 (1999)
4. Bukovinszky, T., Poelman, E.H., Kamp, A., Hemerik, L., Georgios, P., Dicke, M.: Plants under multiple herbivory: consequences for parasitoid search behaviour and foraging efficiency. Animal Behaviour (83), 501–509 (2012)
5. Comins, H.N., Hassell, M.P., May, R.M.: The spatial dynamics of host-parasitoid systems. Journal of Animal Ecology **61**(3), 735–748 (1992)
6. De Bach, P.: The importance of host-feeding by adult parasites in the reduction of host populations. J Econ Entomol **36**(5), 647–653 (1943)
7. Devaney, R.: An introduction to chaotic dynamical systems. CRC Press (2018)
8. Dieckhoff, C., Heimpel, G.E.: Determinants of egg load in the soybean aphid parasitoid Binodoxys communis. Entomologia Experimentalis et Applicata **136**(3), 254–261 (2010)
9. Driessen, G., Bernstein, C.: Patch departure mechanisms and optimal host exploitation in an insect parasitoid. Journal of Animal Ecology **68**(3), 445–459 (1999)
10. Elaydi, S.: An Introduction to Difference Equations. Springer, New York, NY (1996)
11. Emerick, B.: The mathematics of host-parasitoid population dynamics. URL https://www.github.com/bemer42/FURM-Host-Parasitoid-Dynamics
12. Emerick, B., Singh, A.: The effects of host-feeding on stability of discrete-time host-parasitoid population dynamic models. Mathematical Biosciences **272**, 54–63 (2016)
13. Emerick, B., Singh, A., Chhetri, S.R.: Global redistribution and local migration in semi-discrete host-parasitoid population dynamic models. Mathematical Biosciences (327) (2020)
14. Godfray, H.C.J.: Parasitoids; Behavioral and Evolutionary Ecology. Princeton University Press, 41 William St, Princeton, NJ 08540 (1994)
15. Hajek, A.E.: Natural Enemies: An Introduction to Biological Control. Cambridge University Press (2018)
16. Hassell, M.P.: Host-parasitoid population dynamics. J Anim Ecol **69**(4), 543–566 (2000)
17. Hassell, M.P.: The Spatial and Temporal Dynamics of Host Parasitoid Interactions. Oxford University Press, New York, NY (2000)
18. Hassell, M.P., Comins, H.N., May, R.M.: Spatial structure and chaos in insect population dynamics. Nature **353**(6341), 255–258 (1991)
19. Hawkins, B.A.: Patterns and Process in Host-Parasitoid Interactions. Cambridge University Press, New York, NY (1994)
20. Hochberg, M.E., Ives, A.R.: Parasitoid Population Biology. Princeton University Press, Princeton, NJ (2000)
21. Hone, A.N., Irle, M.V., Thurura, G.W.: On the Neimark-Sacker bifurcation in a discrete predatory-prey system. Journal of Biological Dynamics **4**(6), 594–606 (2010)
22. IEEE Conference on Decision and Control: Hybrid systems modeling of ecological population dynamics (2020)
23. Jervis, M.A., Kidd, N.A.C.: Host-feeding strategies in hymenopteran parasitoids. Biol Rev **61**(4), 395–434 (1986)
24. Kidd, N.A.C., Jeris, M.A.: The effects of host-feeding behaviour on the dynamics of parasitoid-host interactions, and the implications for biological control. Res Popul Ecol **31**(2), 2435–274 (1989)
25. Kidd, N.A.C., Jervis, M.A.: Host-feeding and oviposition strategies of parasitoids in relation to host stage: consequences for parasitoid-host population dynamics. Res Popul Ecol **33**(1), 87–99 (1991b)

26. Loke, W.H., Ashley, T.R.: Behavorial and biological responses of Cotesia marginiventris to kairomones of the fall armyworm, Spodoptera frugiperda. Journal of Chemical Ecology **10**, 521–529 (1984)
27. Murdoch, W.W., Briggs, C.J., Nisbet, R.M.: Consumer-Resource Dynamics. Princeton University Press, Princeton, NJ (2003)
28. Murdoch, W.W., Nisbet, R.M., Luck, R.F., Godfray, H.C.J., Gurney, W.S.C.: Size-selective sex-allocation and host-feeding in a parasitoid-host model. J Anim Ecol **61**(3), 533–541 (1992)
29. Nicholson, A., Bailey, V.A.: The balance of animal populations. part 1. Prc Zool Soc London **105**(3), 551–598 (1935)
30. Outreman, Y., Ralec, A.L., Wajnberg, E., Pierre, J.S.: Effects of within- and among-patch experiences on the patch-leaving decision rules in an insect parasitoid. Behavioral Ecology Sociobiology **58**(208-217) (2005)
31. Quicke, D.L.J.: Parasitic Wasps. Chapman and Hall, London (1997)
32. Reeve, J.D., Murdoch, W.W.: Aggregation by parasitoids in the successful control of the California red scale: a test of theory. J Anim Ecol **54**(3), 797–816 (1985)
33. Reigada, C., Araujo, S.B.L., de Aguiar, M.A.M.: Patch exploration strategies of parasitoids: The role of sex ratio and forager's interference in structuring metapopulations. Ecological Modelling (230), 11–21 (2012)
34. Ricker, W.E.: Stock and recruitment. Journal of the Fisheries Research Board of Canada **11**(5), 559–623 (1954)
35. Scheinerman, E.R.: Invitation to dynamical systems. Courier Corporation (2012)
36. Shea, K., Nisbet, R.M., Murdoch, W.W., Yoo, H.J.S.: The effect of egg limitation on stability in insect host-parasitoid population models. J Anim Ecol **65**(6), 743–755 (1996)
37. Singh, A., Nisbet, R.M.: Semi-discrete host-parasitoid models. J Theor Biol **247**(4), 733–742 (2007)
38. Sisson, P., Szarvas, T.: Calculus with Early Transcendentals. Hawkes Learning/Quant Systems, Inc. (2016)
39. Strang, G.: Introduction to Linear Algebra, 5th edn. Wellesley - Cambridge Press (2016)
40. Waage, J., Greathead, D.: Insect Parasitoids. Academic Press (1986)
41. Wajnberg, E., Bernstein, C., van Alphen J.: Behavioral Ecology of Insect Parasitoids: From Theoretical Approaches to Field Applications. Blackwell Publishing, Malden, MA (2008)
42. Wajnberg, E., Colazza, S.: Chemical Ecology of Insect Parasitoids. Wiley Blackwell, Oxford (2013)

Mathematical Modeling of Weather and Climate

Terrance Pendleton

Abstract

Started as a branch of mathematical physics and scientific computing, contemporary applied mathematics has grown significantly in the last several years to involve other disciplines such as weather and climate. This chapter is a focused introduction to mathematical modeling as it relates to climate change and numerical weather prediction. The aims of this chapter are to seek connections between mathematics and physical systems, explain what it means to construct a mathematical model of some real-world phenomena, and exemplify the value of mathematics in problem solving. Moreover, we seek to understand Earth's climate system and predict its behavior through mathematical models, computational experiments, and data analysis.

Suggested Prerequisites *An introductory course in calculus and statistics.*

1   Introduction

How can we use mathematics to help us understand Earth's climate? Better yet, what do we even mean by climate, and how is it different than weather? Weather can be thought of as a mix of events that happen each day in our atmosphere. On the other hand, climate can be thought of as the long-term atmospheric conditions in a region. That is, "climate is what we expect, weather is what we get." (see [6]). Earth's

T. Pendleton (✉)
Drake University, Des Moines, IA, USA
e-mail: terrance.pendleton@drake.edu

climate includes interactions among the atmosphere, hydrosphere (oceans, lakes, and other bodies of water), geosphere (land surface), biosphere (all living things), and cryosphere (snow and ice). The climate system is the exchange of energies and moisture between these spheres.

Given the complexity of Earth's climate, it is no surprise that understanding Earth's climate relies heavily on mathematics. It is not possible for any one person to perceive Earth's climate. Moreover, climate change, and its effects on society and the environment, is one of the most important issues we are currently facing. The evidence of climate change is well known and includes such phenomena as global warning (the gradual increase of Earth's average temperature), loss of ice and sea level rise, and increases in the number of extreme events. Climate scientists use mathematics to (1) describe climate change, (2) predict climate change, and (3) communicate climate change.

1. Climate scientists use mathematics to describe climate change. To see changes in the climate, we must describe the climate in different places and over a period of time. It is for this reason that climate is defined as the statistics of weather. That is, scientists use statistics to try and determine what is normal and what is changing with regards to the climate. For instance, some of the typical quantities measured in climate include temperature, rainfall, humidity, sea level, CO_2 levels, and wind speed. These measurements can be examined using standard statistical techniques (such as *average* rainfall or *maximum* humidity).
2. Climate scientists use mathematics to predict climate change. Mathematical models are typically formulated with the goal of predicting the future behavior of a certain phenomenon. This is the principal tool used by climate scientists. Broadly speaking, a mathematical model is a mathematical representation of the relationship between two or more variables relevant to a given situation or problem. Climate models are used in computational simulations to explore the behavior of the system under various forcing scenarios. These models can be used to determine causes of Earth's climate change. Climate models are typically quite complex and are usually based on multiple sub-models. These models include information on the atmosphere, cloud formation, the oceans, the cryosphere, land surfaces, and so on. The combination of different models requires the modeling of the interaction between different parts of the climate system. For instance, one can model the interaction between the sea levels and rainfall.
3. Climate scientists use mathematics to communicate climate change. Climate science needs to be communicated in order to enact policy change. Thus, it is important not only to communicate these results to the scientific community but also to policymakers and planners in areas such as agriculture, construction, and insurance. This communication requires mathematical literacy, both in the production and consumption of information about climate change for these different audiences. This information can be presented in several forms such as charts, tables, and graphs.

When constructing mathematical models, it is useful to identify broad categories of models. One such division between models is based on the type of outcome they predict. In a *deterministic model*, the output of the model is fully determined by the *parameter values* (arbitrary constants whose values characterize a member of a model), and the *initial conditions* (the starting points or the initial states of the variable). *Stochastic models* possess some inherent randomness. The same set of parameter values and initial conditions will lead to an ensemble of different outputs. Both types of models are used extensively in climate modeling. For instance, modeling random fluctuations of temperature change over a long period of time may be best handled by a stochastic model. On the other hand, modeling the decay of tropical cyclone windspeeds after making landfall can be handled via a deterministic model.

2   Modeling with Data

Data is a powerful tool which can help us translate everyday situations into mathematical statements (or models) which can then by analyzed, validated, and interpreted in context. Depending on the data available, we can use it to help identify assumptions which are consistent with the context of the problem and which in turn shape and define the mathematical characterization of the problem. In this section, we begin by considering one of the simplest mathematical models (linear) and how data supports the development of these kinds of model. From there, we discuss methods for which one can transform nonlinear models into linear ones. Finally, we provide examples of the roles that data play in the development of linear models in the context of climate and weather.

2.1   Linear Models

One of the simplest mathematical models is a linear one which represents the relation between two variables by a straight-line graph. In this scenario, the relationship between two quantities shows a constant rate of change. For instance, consider the following scenario:

Example 1 The city of Mathville has been growing at a constant rate over the last several years. In 2015, the population was 4700. By 2020, the population grew to 6200. Assuming that this trend continues,

1. Predict the population in 2025.
2. Identify the year in which the population will reach 20,000.

The two changing quantities are the size of the population and time. Let us assign the size of the population P or since the population changes as a function of time, $P(t)$, the population of Mathville at time t. To make the computation a little nicer, let us set $t = 0$ to correspond to the year 2015—the earliest year for which we have data. Then the year 2020 will correspond to $t = 5$. To predict the population in 2020 ($t = 5$), we will write an expression for the population as a function of time. Since we are assuming a constant rate of change, we can construct a linear model, which relates the size of the population and the time, of the form

$$P(t) = m \cdot t + b, \tag{1}$$

where m is the rate of change (or slope) and b is the initial size of the population corresponding to $t = 0$ (or the y-intercept). To determine the rate of change, we will use the change in output per change in input. Here the change in input (time) $= 5 - 0$ and the change in output (population size) $= 6200 - 4700$ which corresponds to the ordered pairs: $(0, 4700)$ and $(5, 6200)$. This gives the following rate of change:

$$m = \frac{\text{change in output}}{\text{change in input}} = \frac{6200 - 4700}{5 - 0} = 300. \tag{2}$$

We already know the y-intercept of our line (i.e., $b = 4700$), so we can immediately write the equation:

$$P(t) = 300t + 4700. \tag{3}$$

Using (3) we can predict the population in 2025 by evaluating the linear model at $t = 10$ (recalling that $t = 0$ corresponds to the year 2015). So, the predicted population in 2025 is $P(10) = 300(10) + 4700 = 9700$. To identify the year for which the predicted population is 20,000, we can set $P(t) = 20,000$ and solve for the corresponding t. In doing this, we obtain:

$$20000 = 300t + 4700$$
$$20000 - 4700 = 300t$$
$$\frac{15300}{300} = t$$
$$51 = t.$$

Thus, in 51 years after 2015 (i.e., the year 2066), the predicted population will be 20,000 (Fig. 1).

Exercise 1 In the 1880s, the global mean temperature was 56.71 °F. In the 1930s, the global mean temperature was 56.82 °F. If this trend continues and we assume a constant rate of change,

1. What will be the predicted global mean temperature average in the 1960s?
2. What will be the predicted global mean temperature average in the 2000s?
3. In what decade will the global mean temperature be 70 °F?

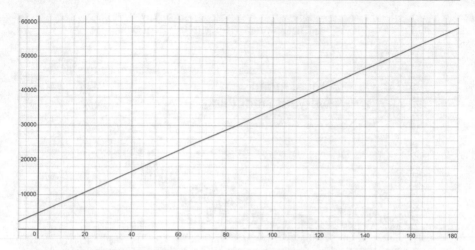

Fig. 1 A graph of $P(t)$ where the x-axis is time (t) measured in years since 2015 and the y-axis is the corresponding population ($P(t)$)

Table 1 Global mean annual temperature average per decade

Decade	°F
1880s	56.71
1890s	56.73
1900s	56.75
1910s	56.76
1920s	56.78
1930s	56.82
1940s	56.85
1950s	56.89
1960s	56.94
1970s	57.03
1980s	57.14
1990s	57.29

Challenge Problem 1 Table 1 containing the global mean annual temperature average per decade over the last several decades.

Notice that the rate of change is **not** constant as it was assumed in the previous exercise. Determine a linear model which predicts the global mean annual temperature average for some decade using the *average* rate of change. Compare your answer to Exercise 1.

Exercise 2 The temperature data after 1960, see Fig. 2, can be approximated by the function

$$T(t) = 0.1t + 0.05, \qquad (4)$$

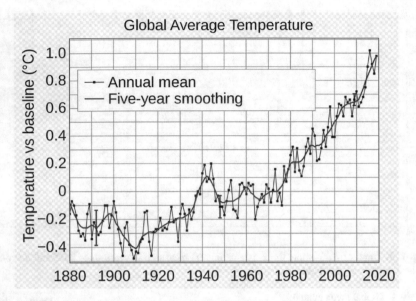

Fig. 2   Global average temperature increase since 1880

where $T(t)$ gives the change in world temperature in degrees Celsius, and t is the number of decades since 1960. For instance, the decade 1960–1969 corresponds to $t = 0$, and at this time, $T(0) = 0.05$. We can interpret this to mean that the world was 0.05 °C warmer than its average temperature before 1960. Graph this function over the period from 1960 to 2050. What does it predict for the temperature change in 2000 and in 2050?

2.2   Transformation to Linear Models

There are situations where it is useful to *transform* variables so that the relationship resembles a straight-line graph. A common method of transformation is the use of logarithms. For instance, suppose that we are told that two variables x and y are related by the exponential function $y = a \cdot b^x$. The graph of this relation is given in Fig. 3. If we take logarithms (of base 10) of each side, we obtain

$$\log y = \log (a \cdot b^x) \implies$$

$$\log y = \log a + \log b^x \implies$$

$$\log y = \log a + x \cdot \log b,$$

Fig. 3 Graph of $y = a \cdot b^x$ against x. Here $a = 2$ and $b = 1.5$

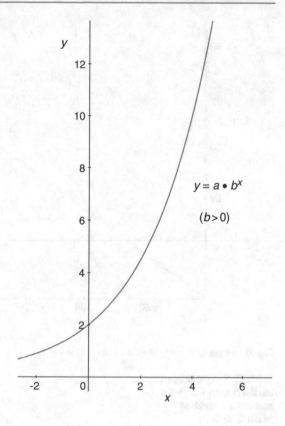

and the graph of $\log y$ against x is a straight line with the slope $m = \log b$ and the y-intercept $b = \log a$ (see Fig. 4.) In the following example, we show how this transformation can be useful in developing a linear model.

Example 2 Let us try and model climate change by modeling the temperature rise of the Earth since the 1870 when the global mean temperature was 56.69 °F. This temperature rise is sometimes referred to as the *greenhouse effect*. This effect is the trapping of the sun's warmth in Earth's lower atmosphere due to the greater transparency of the atmosphere to visible radiation from the sun than to infrared radiation emitted from the planet's surface. This process makes Earth much warmer than it would be without an atmosphere. Gases in the atmosphere, such as carbon dioxide, trap heat like a glass roof of a greenhouse. Because the concentration of carbon dioxide is gradually increasing, this has led to a rise in the average temperature of earth (see [6]). Table 2 shows this temperature rise over the last several decades.

If the average temperature of the Earth rises too high, it could have devastating effects on the climate. As the polar ice caps melt, there could be massive floods which could cause several coastal cities to become completely submerged. In this

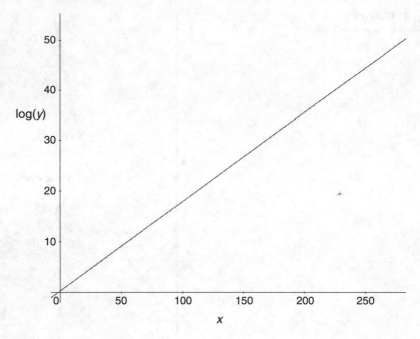

Fig. 4 Graph of log y against x where $y = a \cdot b^x$

Table 2 Temperature rise of the Earth since the global mean temperature was 56.69 °F in 1870

Decade	°F
1880s	0.02
1890s	0.02
1900s	0.04
1910s	0.05
1920s	0.07
1930s	0.11
1940s	0.14
1950s	0.18
1960s	0.23
1970s	0.32
1980s	0.43
1990s	0.58

example, we want to find a model of the data provided in Table 2 and use it to predict when the earth's temperature will be 12 °F warmer than the temperature in 1870. At this temperature, the consequences of global warming will be quite severe.

Let T be the temperature rise of the Earth above the temperature of Earth at 1870 and t be the number of decades since 1870 (so $t = 0$ corresponds to the year 1870 and $t = 1$ corresponds to the year 1880). Without the assistance of data, there is no obvious way of relating the temperature change T with the time t. There are many

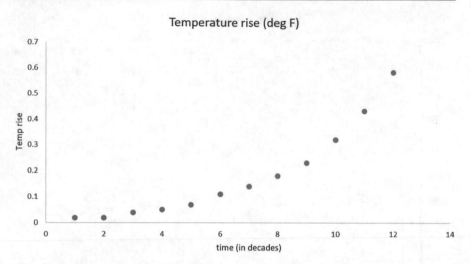

Fig. 5 Graph of T against t from Table 2

complicated processes going on in the atmosphere, and as such there are several physical laws and chemical reactions related to the burning of fossil fuels that must be considered. Because there is no obvious way of discovering the relationship between T and t, we can graph the data (in the form (t, T)) to try and discern some relationship between the temperature and the year. Indeed if we consult Fig. 5, we see that relationship seems to follow that of an exponential function of the form $T(t) = a \cdot b^t$ where a and b must be determined. To identify parameters a and b, we can transform our model to a linear one with the use of logarithms. Recalling that $T = a \cdot b^t$ is equivalent to $\log T = \log a + t \cdot \log b$, we can plot $\log T$ against t to observe the linear relation in Fig. 6. We see that we obtain a straight line through *most* of the data. The points at the lower end do not fit this straight line very well. One possible explanation is that the data is only correct up to two decimal places and so the measurements are not exact. Since we are assuming a linear relationship between $\log T$ and t, we know that the rate of change is *constant* and hence to obtain a linear model, all one has to do is compute the slope by picking two points on the graph given by Fig. 6 (Table 3).

Exercise 3 Determine a linear model of the form $\log T = \log a + t \cdot \log b$ using the ordered pairs $(6, -0.959)$ and $(7, -0.854)$.

Exercise 4 Use the model obtained in Exercise 3 to determine when the Earth's temperature will be 12 °F warmer than its mean temperature in 1870.

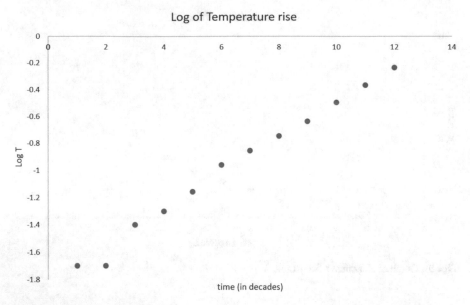

Fig. 6 Graph of $\log T$ against t from Table 3

Table 3 Transformation of the data in Table 2 to a linear model

Decade	t	T	$\log T$
1880s	1	0.02	−1.699
1890s	2	0.02	−1.699
1900s	3	0.04	−1.398
1910s	4	0.05	−1.301
1920s	5	0.07	−1.155
1930s	6	0.11	−0.959
1940s	7	0.14	−0.854
1950s	8	0.18	−0.745
1960s	9	0.23	−0.638
1970s	10	0.32	−0.495
1980s	11	0.43	−0.367
1990s	12	0.58	−0.237

Exercise 5 Repeat Exercise 3 using the pair $(8, -0.745)$ and $(9, -0.638)$ and compare your results to Exercise 3.

Challenge Problem 2 Notice that the rate of change is **not** constant as it was assumed in the previous exercises. Determine a linear model of the form $\log T = \log a + t \cdot \log b$ which predicts the global mean annual temperature average for some decade using the *average* rate of change. Compare your answer to Exercises 3 and 5.

3   The Keeling Curve

The Keeling Curve is a graph of the accumulation of carbon dioxide in the Earth's atmosphere which is based on continuous measurements taken at the Mauna Loa Observatory on the island of Hawaii from 1958 to present day. It provides one of the clearest indications of the possibility of global warming.

In the nineteenth century scientists had established that carbon dioxide in the atmosphere warms the planet—creating the *greenhouse effect*. The greenhouse effect is essential for life on Earth; without it, Earth would be a frigid place devoid of life. However, it was speculated that burning fossil fuels might increase CO_2 levels which in turn causes the global temperatures to increase. Keeling's work established that burning fossil fuels was indeed causing an increase in carbon dioxide in the atmosphere.

The clear and vivid illustration provided by the Keeling Curve spurred other researchers to begin looking at the possible impacts. By 1967, a team led by Syukuro Manabe of the Geophysical Fluid Dynamics Laboratory of the National Oceanic and Atmospheric Administration (NOAA) had devised the first comprehensive model of the response of climate to an increase in atmospheric CO_2 extrapolated from the Keeling Curve. It predicted that a doubling of carbon dioxide in the atmosphere would cause an increase in global temperature of around 3–4 °F (see [19]).

In this section, we will work with climate data in the form of the Keeling Curve to develop models that allow us to make predictions regarding the increase in CO_2 levels for future years. We will use MATLAB to assist us in performing these calculations and producing plots. However, this kind of analysis could be performed on any standard statistical package. To develop a linear model of the data graphed in Fig. 7, we rely on a statistical method known as *linear regression*.

3.1   Linear Regression

Suppose that we have a set of data points $\{(x_i, y_i) : i = 1, \ldots, n\}$. If we plot the data, we might be able to hypothesize some relationship between x and y (such as a linear or exponential relationship). How can we use *all* of the data points to determine a *best* fit? If we assume a linear relationship, we know that the rate of change is constant; however, because the data may not be perfectly linear (and it usually is not!) any two points that we choose to compute the rate of change will yield different results. In the previous exercises, we explore one way of using data points to mathematically characterize the linear relationship between x and y. The goal of linear regression analysis is to find a functional relationship between x and y that best approximates the data. In other words, we seek a function $f(x)$ so that $y_i \approx f(x_i)$ for each data point where $f(x_i) = mx_i + b$.

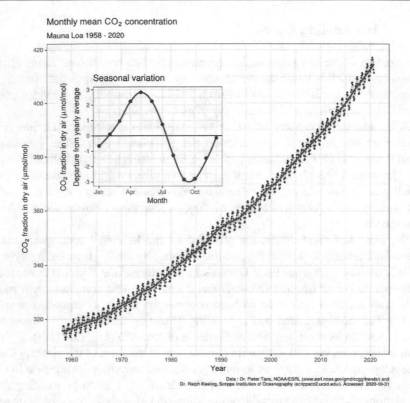

Fig. 7 Measurements of atmospheric carbon dioxide levels from Mauna Loa observatory

3.1.1   The Method of Least Squares

Suppose that we observe that the relationship between x and y is linear, so that our model takes the form $y = mx + b$. As we saw in Chapter 2, the goal is to find values for m and b that provide a good approximation of the linear relationship. Ideally, we would find these values so that

$$y_i = f(x_i) = mx_i + b. \tag{5}$$

As we do not expect to find a constant value m and b so that this equality is maintained for each $i = 1, 2, \ldots, n$, we can formulate this problem as finding values for m and b such that the *error* $|y_i - f(x_i)|$ is as small as possible. That is, we try and minimize the distance between y_i and $f(x_i)$. Here, y_i corresponds to the actual data point; whereas, $f(x_i)$ corresponds to the predicted value determined by our linear model. Notice that in this formulation, we are now using every data point. If we were to find m and b such that Eq. (5) was satisfied for each data point, we would generate the following system of equations

$$mx_1 + b = y_1$$
$$mx_2 + b = y_2$$
$$\vdots$$
$$mx_n + b = y_n. \tag{6}$$

Since this is n equations with two unknown quantities (m and b), this system is said to be *overdetermined* and will typically not have a solution. This, of course, is no surprise to us since the system of equations above assumes that we can perfectly fit the data into a linear equation. The goal then, is to find the *best fit* to the data. We can express the system of equations above in a more compact form using matrices. For example, if we are given the system of equations

$$2x_1 + 3x_2 = 8$$
$$5x_1 + 7x_2 = 14, \tag{7}$$

then we can let $A = \begin{bmatrix} 2 & 3 \\ 5 & 7 \end{bmatrix}$ be the *coefficient matrix*, $x = \begin{bmatrix} x_1 \\ x_2 \end{bmatrix}$ be the matrix corresponding to the unknown variables, and $b = \begin{bmatrix} 8 \\ 14 \end{bmatrix}$ be the matrix corresponding to the right-hand sides of the equations. Then Eq. (7) can be expressed in matrix form as $Ax = b$ or

$$\begin{bmatrix} 2 & 3 \\ 5 & 7 \end{bmatrix} \begin{bmatrix} x_1 \\ x_2 \end{bmatrix} = \begin{bmatrix} 8 \\ 14 \end{bmatrix}. \tag{8}$$

In a similar manner, we can express (6) as $Ax = Y$, where

$$A = \begin{bmatrix} x_1 & 1 \\ x_2 & 1 \\ \vdots & \vdots \\ x_n & 1 \end{bmatrix}, \quad x = \begin{bmatrix} m \\ b \end{bmatrix}, \quad Y = \begin{bmatrix} y_1 \\ y_2 \\ \vdots \\ y_n \end{bmatrix}. \tag{9}$$

The values of A and Y can be determined from the data, but the matrix x is unknown. We would like to make $|y_i - f(x_i)|$ as small as possible, and this is equivalent to trying to make the entries of the matrix $Y - Ax$ as small as possible. In other words, we want to minimize the length of the error matrix $Y - Ax$. This can be achieved by minimizing the following quantity:

$$(y_1 - (mx_1 + b))^2 + (y_2 - (mx_2 + b))^2 + \ldots + (y_n - (mx_n + b))^2. \tag{10}$$

Minimizing the sum of the squares of the errors is called the *method of least squares*.

The *length* of a vector $x = \begin{bmatrix} x_1 \\ \vdots \\ x_n \end{bmatrix}$ (a single column (or row) matrix), denoted by $\|x\|$,

is given by

$$\|x\| = \sqrt{x_1^2 + x_2^2 + \ldots + x_n^2}, \tag{11}$$

and hence we seek to minimize $\|Y - Ax\|^2$. The least squares solution to $Ax = Y$ is the vector x satisfying

$$A^T A x = A^T Y, \tag{12}$$

where $A^T = \begin{bmatrix} x_1 & x_2 \ldots & x_n \\ 1 & 1 \ldots & 1 \end{bmatrix}$. This is called the *transpose* of A. If A is an $n \times p$ matrix (i.e., n data points and p unknown values in the formula for f), then $A^T A$ is a $p \times p$ symmetric matrix (i.e., $S^T = S$). If the columns of A are linearly independent (i.e., if none of the columns can be expressed as a linear combination of the other columns of A), which will be the case for applications that we study in this chapter, then $A^T A$ is invertible and the least squares solution is uniquely given by

$$x = (A^T A)^{-1} A^T Y, \tag{13}$$

where $(A^T A)^{-1}$ denotes the *inverse* of $A^T A$. This solution can be quickly computed in MATLAB and will give us values for m and b, assuming that we are using linear regression.

3.1.2 The Coefficient of Determination

Once we obtain our least squares solution in (13), we have the best fit for $f(x) = mx + b$. We define the *residuals* as the measurements of how far each data point is from the fit. Mathematically, this is given by $r_i = y_i - (mx_i + b)$. In a more compact form, we can define the *residual vector r* as $r = Y - Ax$. The length of r allows us to determine the quality of our fit. A commonly used measure in statistics is the *coefficient of determination* or R^2 value. If we let $\bar{y}$ denote the mean of the y_i data, that is,

$$\bar{y} = \frac{1}{n} \sum_{i=1}^{n} y_i,$$

and $\bar{Y}$ be the $n \times 1$ vector

Table 4 A list of useful Matlab commands

Command	Definition
inv(A)	Gives the inverse of an $n \times n$ matrix A
A'	Determines the transpose of an $n \times n$ matrix A
A*B	Determines the matrix product $A \cdot B$
A*b	Determines the matrix-vector product $A \cdot b$
a = [1:1:100]	Creates a vector with equally spaced entries from 1 to 100
a.^2	Squares each entry in the vector a
a(2:5)	Gives the 2nd through 5th entries in the vector a
norm(a,2)	Gives the length of a vector a
a(1)	Gives the 1st entry in a vector a
A = [a b]	Creates a matrix A with the columns a and b
mean(a)	Determines the mean of the entries in a vector a
ones(10,1)	Creates a vector with 10 ones
plot(x,y,'b')	Generates a plot of the points in vectors x and y in blue
plot(x,y,'b',w,z,'r')	Generates two plots on the same coordinate plane in different colors

$$\bar{Y} = \begin{bmatrix} \bar{y} \\ \bar{y} \\ \vdots \\ \bar{y} \end{bmatrix},$$

then the coefficient of determination is given by the formula

$$R^2 = \frac{\|Ax - \bar{Y}\|}{\|Y - \bar{Y}\|}. \tag{14}$$

The value of R^2 measures how well the fit Ax does in comparison to the mean $\bar{Y}$. The closer to 1, the better the fit.

Example 3 In this example, we show how one can use computational tools such as MATLAB to perform a regression analysis on a set of data. Table 4 is a useful table of MATLAB commands that we will make use of.

Consider the following set of data points and its corresponding plot (Table 5 and Fig. 8). Observe that the plot suggests a linear relationship between x and y. We perform a linear regression analysis to find a linear model which best fits the data points.

Observing from Fig. 8 that we have a linear relationship, we can perform a linear regression analysis as follows:

Table 5 Sample data points,
for Example, 3

x	y
1	11.8
2	20.4
3	23.5
4	19.1
5	31.1
6	30.9
7	34.2
8	39.2
9	42.7
10	41.4
11	48.2
12	48.4

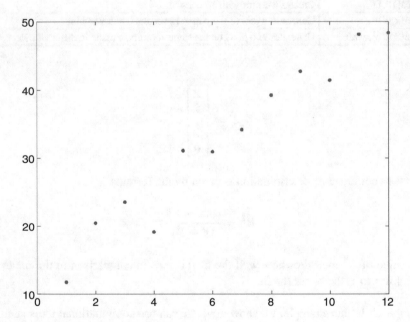

Fig. 8 A plot of the data from Table 5

1. Enter the data into two column vectors *xdata* and *ydata*.

```
xdata=[1:1:12]'
ydata=[11.8 20.4 23.5 19.1 31.1
30.9 34.2 39.2 42.7 41.4 48.2 48.4]'
```

2. Form the matrix A given in Eq. (9).

```
A = [xdata ones(12,1)]
```

3. Fit the data to a straight line using the method of least squares and Eq. (13).

```
x=inv(A'*A)*A'*ydata
x =
3.2297
11.5818
```

Here we obtain that $m = 3.2297$ and $b = 11.5818$. Thus our best fit line is given by $y = 3.23x + 11.58$.

4. Using Eq. (14), find the R^2 value.

```
ybar=mean(ydata)*ones(12,1)
r2top=norm(A*x - ybar, 2)      %this is the numerator of
  r2
r2bottom=norm(ydata-ybar,2)    %this is the denominator
  of r2
r2 = r2top/r2bottom
```

Here, we obtain $R^2 = \frac{38.622}{39.626} \approx 0.975$ which indicates a good fit as this is a value close to 1. We can confirm this visually with a plot. To generate a plot of the data and the best fit line, we use the following commands in MATLAB:

```
yfit = x(1)*xdata + x(2)
plot(xdata,ydata,'.',xdata,yfit,'r'),
```

recalling that $x(1) = m$ and $x(2) = b$ for our best fit line $f(x) = mx + b$. In Fig. 9 we obtain a visual confirmation of our fit.

Exercise 6 In this exercise, we perform a linear regression analysis on the data obtained from the measurements of carbon dioxide (CO_2) at the Mauna Loa Observatory at the Mauna Loa volcano in Hawaii which was begun by Charles Keeling in March of 1958 and continue today. Air samples have been taken hourly, every day, using the same measuring techniques from the very beginning.

1. Begin by downloading the data set (which contains measurements from March 1958 to February 2018) *C02Data-MLO.xlsx* via the link:
   https://1drv.ms/x/s!Aq7IzuMtCFE8nVfvpZaQcRI_wHlh?e=Tj0WEu
   To import the data into MATLAB, click on the Home tab, and then the Import Data button (which is a green arrow). This will open up a new window to select the correct spreadsheet. We want to import the relevant data as (column) vectors in MATLAB. The columns to import are column C (the decimal dates), Column E (the interpolated data), and column F (the seasonal corrections). You may need to adjust the range of your imported data to exclude

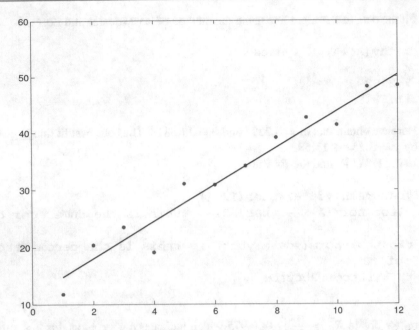

Fig. 9 A plot of the data from Table 5 and the best fitting line obtained from the method of least squares

the first couple of rows. Click on the `Import Selection` button and then choose `Import Data` from the drop-down menu. This should create a vector in MATLAB that is visible in your workspace. Change the name of the vector by clicking on the vector name in the workspace.

2. Make a plot of the Keeling Curve including both the interpolated data and seasonal corrections. To replicate the Keeling curve in Fig. 7, the x-axis should be the decimal date and the y-axis should be the CO_2 concentrations.
3. Explain the usefulness of decimal dates. For instance, why is July of 1958 written as 1958.542?
4. Fit the interpolated data to a straight line using the method of least squares and Eq. (13). Find the coefficient of determination and comment on the quality of the fit. According to the fit, what is the average predicted increase in CO_2 per year?
5. Use the fit to estimate the CO_2 at Mauna Loa in January of 2030, 2050, 2070, and 2100.

Challenge Problem 3 One can extend the concept of the method of least squares to determine a quadratic model of the form $f(w) = a_0 + a_1 w + a_2 w^2$ by adjusting the matrices A and x as follows:

$$A = \begin{bmatrix} x_1^2 & x_1 & 1 \\ x_2^2 & x_2 & 1 \\ \vdots & \vdots & \vdots \\ x_n^2 & x_n & 1 \end{bmatrix}, x = \begin{bmatrix} a_2 \\ a_1 \\ a_0 \end{bmatrix}. \tag{15}$$

Fit the data to a quadratic function using the method of least squares. Plot the original data along with the quadratic fit. Find the coefficient of determination and comment on the quality of the fit. Compare the new result with the linear approximation obtained in the previous exercise. Which approximation is better?

3.2   Modeling with Differential Equations

Another approach for modeling the increase of atmospheric carbon dioxide involves the use of *differential equations*. A differential equation is an equation that relates the rate, $\frac{dy}{dt}$, at which a quantity y is changing (or sometimes a higher derivative) to some function $f(t, y)$ of that quantity and time.

Example 4

$$\frac{dy}{dt} = 5, \quad \frac{dx}{dt} = 5x - 3, \quad y^{-2}\frac{dy}{dx} = x^4$$

are all examples of *first-order* differential equations. The order denotes the highest order derivative that appears in the differential equation.

Modeling with differential equations relies on making assumptions about the rate at which a quantity is changing. For instance, suppose that $T(t)$ is a function that gives the temperature of a cup of hot tea that is cooling with respect to time. If we assume that the rate at which the cup of tea cools is proportional to the current temperature of the tea, we could write the differential equation $\frac{dT}{dt} = kT$, where k is the constant of proportionality. This assumption seems valid since we expect the tea to cool faster if it is hotter (and slower if the tea is cooler). Furthermore, since the tea is cooling, we can also assume that $k < 0$. If we wanted to develop a model for the temperature of the cup of tea, we can solve the differential equation above by a process known as *separation of variables*.

Example 5 To find the general solution to $\frac{dT}{dt} = kT$, we can use the process of *separation of variables*. This technique works if we can rearrange a differential equation, $\frac{dy}{dx} = p(y)q(x)$, to have the form:

$$r(y)\, dy = q(x)\, dx,$$

where $r(y) = \frac{1}{p(y)}$. From this form, we integrate both sides of the equation. For a general solution, we add a constant of integration. Using this method, we obtain

$$\frac{dT}{dt} = kT \implies$$

$$dT = kT\,dt \implies$$

$$\frac{dT}{T} = k\,dt \implies$$

$$\int \frac{dT}{T} = \int k\,dt \implies$$

$$\ln|T| = kt + C \implies$$

$$e^{\ln|T|} = e^{kt+C} \implies$$

$$T = e^C e^{kt} = \hat{C} e^{kt},$$

where $\hat{C} = e^C$. Notice that the absolute value sign around T was dropped since the expression that corresponds to T is necessarily nonnegative. To find a value of $\hat{C}$ we need more information about our modeling scenario. For instance, if we denote the initial temperature of the cup of tea as T_0, then $T(0) = \hat{C} e^0 = \hat{C} = T_0$ and so our model is given by $T(t) = T_0 e^{kt}$.

There are many pathways that contribute to the current atmospheric carbon dioxide levels. Currently, there are about 3000 gigatons of carbon dioxide in the atmosphere. Carbon sinks absorb more carbon than they release, whereas carbon sources release more carbon than they absorb. It is worth noting that for every 44 gigatons of the carbon dioxide molecule, there are 12 gigatons of the element carbon (see [20]). Some of the major sources of carbon (signified with a +) are

- Vegetation (+119.6 gigatons/yr)
- Oceans (+88 gigatons/yr)
- Human activity (+6.3 gigatons/yr)
- Changing land use (+1.7 gigatons/yr)

Major sinks of carbon (signified with a -) are

- Vegetation (−120 gigatons/yr)
- Oceans (−90 gigatons/yr)
- Changing land use (−1.9 gigatons/yr)

Exercise 7 From the values of the sources and sinks given above, and assuming that they are constant, create a differential equation that gives the rate of change of atmospheric carbon, $C(t)$, in gigatons.

Exercise 8 Integrate the equation in Exercise 7. Assume that $C(2005) = 730$ gigatons. Recalling that for every 44 gigatons of the carbon dioxide molecule, there are 12 gigatons of the element carbon, derive a model which gives the total amount of carbon dioxide as a function of time.

Exercise 9 What does the model predict for the amount of carbon dioxide in the atmosphere in 2050 if the above source and sink rates remain the same?

Exercise 10 If the current carbon dioxide abundance is 384 ppm, what does the model predict for the carbon dioxide abundance in 2050?

Exercise 11 Does the answer for the net change in Exercise 7 match up with the Keeling Curve data that indicates a net annual increase of carbon dioxide of 11 gigatons/year?

4 Energy Balance Models

In this section, we introduce a way to approach Earth's "climate system" mathematically through a global *energy balance model* (EBM). Energy balance models are climate models that try to predict the average surface temperature of the Earth from solar radiation, emission of radiation to outer space, and Earth's energy absorption and greenhouse effects. A global EBM summarizes the state of the Earth's climate system in a single variable–the temperature at the Earth's surface averaged over the entire globe. We begin with the simplest model and assume that Earth is a homogeneous solid sphere. That is, we ignore differences in topography (altitude), differences in the atmosphere's composition (such as clouds), differences among continents and oceans, and so on. Because we are not considering any spatial variations, these models are sometimes referred to as *zero-dimensional energy balance models*. For more information about energy balance models, we refer the readers to [10–18].

4.1 Observation

The climate system is powered by the sun, which emits radiation in the ultraviolent (UV) regime (wavelength less than $0.4\,\mu\text{m}$). This energy reaches the Earth's surface, where it is converted by physical, chemical, and biological processes to radiation in the infrared (IR) regime (wavelength greater than $5\,\mu\text{m}$). This IR radiation is then re-emitted into space. If the Earth's climate is in *equilibrium*, the average temperature of Earth's surface does not change, so the amount of energy received must equal the amount of energy re-emitted.

To build the model, we need to introduce the following units, variables, and physical parameters.

4.2   Units

- Length: meter (m), a μm is a micrometer $= 10^{-6}$ m, a km is a kilometer $= 10^3$ m.
- Energy: watt (W). 1 watt $= 1$ joule per second $= 1\frac{\text{kg}\cdot\text{m}^2}{\text{sec}^3}$. A joule is the unit of energy used by the International Standard of Units (SI). It is defined as the amount of work done on a body over a distance of one meter.
- Temperature: kelvin (K). An object whose temperature is 0 K has no thermal energy, i.e., 0 K is *absolute zero*. The Kelvin scale is closely related to the Celsius scale. The magnitude of a degree in the Celsius scale is the same as the magnitude of a kelvin in the Kelvin scale, but the zero point is different. Water freezes at the zero point in Celsius and at 273.15 K. Thus, the Kelvin scale is the Celsius scale plus 273.15.

4.3   Variable

- T, the temperature of the Earth's surface averaged over the entire globe.

4.4   Physical Parameters

- R (6378 km), the radius of the Earth.
- S, the energy flux density–the rate of transfer of energy through a surface or rate of energy transfer per unit area. Through satellite observations, $S = 1367.6$ Wm^{-2}.
- σ (sigma), Stefan-Boltzmann constant; its value is given by $\sigma = 5.67 \cdot 10^{-8}$ Wm^{-2} K^{-4}. The Stefan-Boltzmann constant is the constant of proportionality in the Stefan-Boltzmann law: "the total energy radiated per unit surface area of a black body across all wavelengths per unit time is directly proportional to the fourth power of the black body's thermodynamic temperature T." By *black body* we mean a physical body that absorbs all incident electromagnetic radiation.

Next, we make use of the following assumptions:

4.5   Assumptions

- Viewed from the sun, the Earth is a disk.

- The area of the disk as seen by the sun is πR^2.
- Recalling that the energy flux density is S, the amount of energy flowing through the disk (i.e., reaching the Earth) is

$$\text{Incoming energy (W)}: E_{in} = \pi R^2 S. \tag{16}$$

- All bodies radiate energy in the form of electromagnetic radiation.
- In physics, it is shown that for "black-body radiation" the temperature dependence is given by the Stefan–Boltzmann law (see above, in units of Wm^{-2}),

$$F_{SB}(T) = \sigma T^4. \tag{17}$$

- The area of the Earth's surface is $4\pi R^2$.
- The amount of energy radiated out by the Earth is

$$\text{Outgoing energy (W)}: E_{out} = 4\pi R^2 \sigma T^4. \tag{18}$$

Exercise 12 Recalling that energy is measured in watts, verify that the units of E_{in} and E_{out} are in watts (W). Note that if the incoming energy is greater than the outgoing energy, the Earth's temperature will increase. Likewise, if the outgoing energy is greater than the incoming energy, the Earth's temperature will decrease. We are interested in the case where the Earth's temperature remains constant. That is, Earth is in *thermal equilibrium*. Determine the temperature T for which Earth is in thermal equilibrium, i.e., $E_{in} = E_{out}$ (energy balance equation, see Eqs. (16) and (18)). The known average temperature of the Earth is about $16\,°C$ (or $287.7\,K$). How well does the answer compare with the known average temperature?

You may have noticed that your answer in Exercise 12 was pretty far off from $16\,°C$. It turns out that the model posed in Exercise 12 was too simple and omitted too many different factors. For instance, we did not consider the fact that snow, ice, and clouds can reflect a significant amount of incoming energy from the Sun. This fraction of energy that is reflected back into space before it reaches the Earth's surface is called the (planetary) *albedo*. Let α denote the albedo, so that the remaining fraction $1 - \alpha$ (sometimes called the *co-albedo*) of the incoming solar radiation will reach the Earth's surface (Fig. 10).

Exercise 13 Adding the effects of albedo to E_{in}, adjust the energy balance equation and determine the temperature T for which Earth is in thermal equilibrium. A typical value for Earth's *average* albedo is $\alpha = 0.30$. This means that about 70% of the incoming energy is absorbed by the Earth's surface. How well does the answer compare with the known average temperature? Is this new model an improvement over the previous model? Why/Why not?

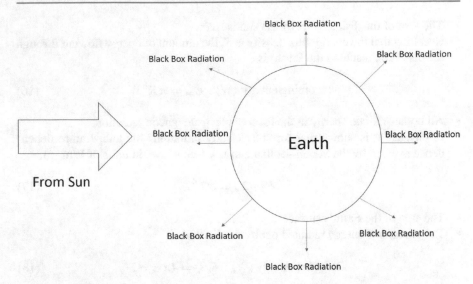

Fig. 10 A conceptual model of Earth's climate system: incoming sunlight and outgoing heat

You may have noticed that your answer in Exercise 13 is even worse than the solution in Exercise 12–even though this new model is more physically relevant! Reminding ourselves that modeling is an iterative process, rather than throw away the albedo introduced in Exercise 13, we add another factor which has a significant effect on the global equilibrium temperature. A good portion of the difference between the answer in Exercise 13 and the average global temperature can be attributed to the *greenhouse effect* of Earth's atmosphere. That is, we include the effects of greenhouse gases (like carbon dioxide and methane), water, dust, and aerosols on the atmosphere. The chemical properties of these greenhouse gases have a significant effect on the atmosphere by reducing the Stefan-Boltzmann law by some factor. This, in turn, affects the outgoing energy.

Let ϵ denote the *greenhouse factor* which is an artificial parameter used to model the effect of greenhouse gases on the permittivity of the atmosphere. While the value of ϵ is unknown, we will assume that $0 < \epsilon < 1$.

Exercise 14 Write a new energy balance equation which incorporates albedo and the greenhouse effect.

Exercise 15 Because the value of ϵ is unknown, it is not possible to determine the equilibrium temperature T. However, we can work backward to determine the value ϵ^* so that the equilibrium temperature is $T = 287.7$ K. Determine this value ϵ^*.

Exercise 16 Suppose that the combined effects of greenhouse gases, dust, and aerosols reduces the parameter ϵ so that $\epsilon < \epsilon^*$. What happens to the equilibrium temperature? Is this what was expected to have happen? *Remark: It may be helpful to read up on the greenhouse effect. See [9], for instance.*

Earlier we mentioned that if the incoming energy is greater than the outgoing energy, the Earth's temperature will increase. Likewise, if the outgoing energy is greater than the incoming energy, the Earth's temperature will decrease. Suppose the temperature is increasing. Will the temperature continue to increase, or will the temperature eventually level off? How fast will the temperature change? To answer these types of questions, we must adjust our model so that it allows the temperature to change over time. Perhaps the simplest model is one that assumes that the temperature changes at a rate proportional to the energy imbalance.

Exercise 17 Rewrite the last sentence as a mathematical equation using E_{in} and E_{out} to obtain a differential equation.

4.6 Phase Line Analysis

A phase line is a diagram that shows the qualitative behavior of a differential equation of the form

$$\frac{dy}{dt} = f(y). \tag{19}$$

This is a useful technique for analyzing the long-term behavior of a differential equation when obtaining a solution is difficult or impossible. To construct a phase line for Eq. (19) we begin by finding the values of y for which y does not change (i.e., $y(t) = C$ for all values of t). This occurs precisely at those values where $\frac{dy}{dt} = 0$ or $f(y) = 0$. These values are referred to as the *critical points* or *equilibrium points* of the differential equation. Outside these critical points, y is either increasing (so that $\frac{dy}{dx} = f(y) > 0$) or y is decreasing (so that $\frac{dy}{dx} = f(y) < 0$). To draw a phase line, we begin with a (usually vertical) line which represents an interval of the domain of the derivative. The intervals between the critical points have their sign indicated with arrows: an interval over which the derivative is positive has an arrow pointing in the positive direction along the line (up), and an interval over which the derivative is negative has an arrow pointing in the negative direction along the line (down). This construction is reminiscent of the first derivative test when finding local minimums and/or maximums.

A critical point can be classified as stable (sink), unstable (source) or semi-stable (node) by inspecting the behavior above and below the critical (i.e., if $f(y) > 0$ or $f(y) < 0$). If both arrows point toward the critical point, it is stable and nearby solutions will converge to this critical point. If both arrows point away from the critical point, it is unstable and nearby solutions will diverge from the critical point. On the other hand, if one arrow points toward the critical point, and one points away,

Fig. 11 Phase line for $\frac{dy}{dx} =$
$-(y-1)(y-3)(y-5)^2(y-7)$

the critical point is said to be semi-stable in that the critical point is stable in one direction and unstable in the other direction.

Example 6 Consider the following differential equation

$$\frac{dy}{dx} = -(y-1)(y-3)(y-5)^2(y-7). \tag{20}$$

To draw a phase line, we find the critical points by setting the right-hand side of Eq. (20) equal to 0 to obtain $y = 1, 3, 5, 7$. To classify these critical points, we divide our phase line into the intervals $(-\infty, 1)$, $(1, 3)$, $(3, 5)$, $(5, 7)$ and $(7, \infty)$. We observe that $\frac{dy}{dx} < 0$ in the intervals $(1, 3)$ and $(7, \infty)$ and $\frac{dy}{dx} > 0$ in the intervals $(-\infty, 1)$, $(3, 5)$, and $(5, 7)$. Consulting Fig. 11, we find that $y = 1$ and $y = 7$ are stable critical points, $y = 3$ is an unstable critical point and $y = 5$ is a semi-stable critical point.

It is traditional to formulate the differential equation of temperature evolution in terms of energy densities (Wm^{-2}). In Exercise 17, we obtained a differential equation expressed in terms of E_{in} and E_{out} which are energies measured in watts (W). To convert these values to energy densities, we can divide by the Earth's surface area (πR^2). In terms of energy densities, the *temperature evolution equation* becomes

$$C\frac{dT}{dt} = \frac{1}{4}(1-\alpha)S - \epsilon\sigma T^4, \tag{21}$$

where C is the *planetary heat capacity* which connects the rate of change of the temperature to energy densities and is the amount of energy needed to raise the temperature of the planet by 1 K. Note that even in this new scenario, if $E_{in} = E_{out}$, the derivative is zero indicating that Earth is in thermal equilibrium.

Exercise 18 With the differential equation given by Eq. (21) (and taking $\alpha = .3, \epsilon = 0.66, C = 1$):

1. Determine the effect of $E_{in} > E_{out}$ on the global average temperature T. Does the answer make sense physically?
2. Determine the effect of $E_{in} < E_{out}$ on the global average temperature T. Does the answer make sense physically?
3. Suppose that the current temperature is 350 K. Is it expected for the temperature to increase, decrease, or remain the same?
4. Suppose that the current temperature is 250 K. Is it expected for the temperature to increase, decrease, or remain the same?
5. What do the answers given above suggest about the *stability* of the equilibrium point? Defend your answer by performing a phase line analysis. Assume that a reasonable domain for T is [200, 400].
6. Now suppose that $\epsilon = 0.5$. Determine the stability of the equilibrium point by performing a phase line analysis. How does this analysis compare to the previous problem?

So far, we have assumed that the albedo is constant and independent of the surface temperature. However, this assumption does not account for the fact that when the surface temperature is sufficiently low, water turns to ice and increases the ability for Earth to reflect incoming energy from the Sun. Thus, we should consider a temperature-dependent albedo with the following constraint:

$$\alpha(T) \approx \begin{cases} 0.7 & T < 250, \\ 0.3 & T > 280. \end{cases} \tag{22}$$

This allows us to incorporate the assumption that when T is low enough, water turns to ice and increases the albedo. We are now in a position to adjust our temperature evolution equation given by Eq. (21) by replacing α with a monotonically decreasing function $\alpha(T)$ that connects the value 0.7 at $T \approx 250$ K with the value 0.3 if $T > 280$ K. There are several ways of accomplishing this. One such way is given by: (See Fig. 12)

$$\alpha(T) = 0.5 - 0.2 \cdot \tanh\left(\frac{T - 265}{10}\right), \tag{23}$$

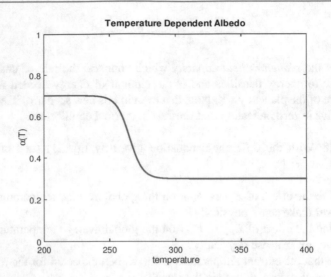

Fig. 12 Graph of Eq. (23)

where $\tanh(x) = \dfrac{e^{2x} - 1}{e^{2x} + 1}$ is the *hyperbolic tangent function.*

Our temperature evolution equation now becomes

$$C\frac{dT}{dt} = \frac{1}{4}(1 - \alpha(T))S - \epsilon\sigma T^4, \tag{24}$$

with $\alpha(T)$ given by Eq. (23). We would like to perform a similar phase line analysis as in the previous problem. Since this is a new differential equation, we begin by finding the equilibrium points. Noticing that the right-hand side of the differential equation is a complicated nonlinear function (of T since all other values are known), we must estimate the equilibrium points using a root finding method. The equilibrium points of the differential equation given by Eq. (24) occur when $E_{in} = E_{out}$ (Why?). Since E_{in} and E_{out} are both functions of time, we can plot E_{in} and E_{out} to see that Eq. (24) has three equilibrium points. (See Fig. 13.) We will use the "Solver" add-in in Excel to find the values of T so that $f(T) = \frac{1}{4}(1 - \alpha(T))S - \epsilon\sigma T^4 = 0$. Before we access Solver, we must first input $f(T)$ into Excel. (See Fig. 14.) In Fig. 14, A1 $= f(200)$.

4.7 Using Excel's Solver

The procedure for using Excel's Solver is as follows:

1. In Excel, Solver can be accessed in one of two ways, depending on which version of Excel is being used. If you are using Excel 2019, under the "Tools" menu select

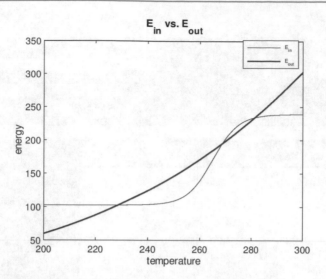

Fig. 13 Graph of $E_{in} = \frac{1}{4}(1 - \alpha(T))S$ vs. $E_{out} = \epsilon\sigma T^4$ with $\alpha(T)$ is given in Eq. (23)

A1		× ✓ f_x	=(0.25*(1-(0.5-0.2*TANH((B1-265)/10)))*1367.6)-(0.66*5.67*10^(-8)*B1^4)							
	A	B	C	D	E	F	G	H	I	J
1	42.69511	200								
2										

Fig. 14 Inputting $f(T)$ into excel

"Solver." A new pop-up window will appear. *Remark*: If this is not an option, the add-in will need to be installed. To access Solver, select "Add-Ins" under the "Tools" menu and check the solver add-in.

Otherwise, if you are using an older version of Excel, you can access Solver in the Analysis group under the "Data" tab. *Remark*: If you do not see this as an option, the add-in will need to be installed. To access Solver, go to File > Options. Click Add-Ins, and then in the Manage box, select Excel Add-ins. Click go. In the Add-Ins available box, select the Solver Add-in check box, and then click OK. After you load the Solver Add-in, the Solver command is available in the Analysis group on the Data tab (see Fig. 15).

2. In the box labeled "Set Objective:" enter the reference of the cell into which we typed the formula. In our example, we would type A1.

3. Click the "Value of:" button. Enter the target value in the "Value Of:" box. In our example, we would type 0. (This is the right-hand side of the nonlinear equation that we are trying to solve.)

4. In the "By Changing Variable Cells:" box, type the formula's reference cell. In our example, we would type B1.

5. Click "Solve." Excel will change both cells accordingly.

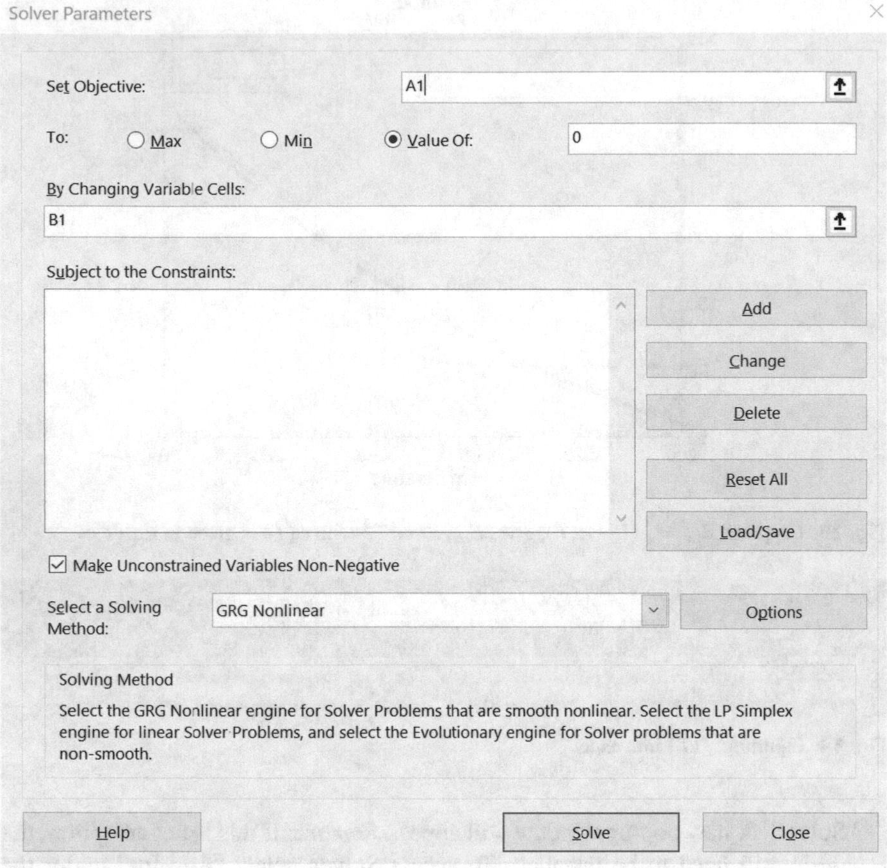

Solver Parameters   ✕

Se_t Objective: A1| ⬆

To: ○ Max   ○ Mi_n   ⦿ _Value Of: 0

_By Changing Variable Cells:

B1   ⬆

S_ubject to the Constraints:

		Add
		_Change
		_Delete
		_Reset All
		_Load/Save

☑ Ma_ke Unconstrained Variables Non-Negative

Select a Solving   GRG Nonlinear   ⌄   Op_tions
Method:

Solving Method

Select the GRG Nonlinear engine for Solver Problems that are smooth nonlinear. Select the LP Simplex
engine for linear Solver Problems, and select the Evolutionary engine for Solver problems that are
non-smooth.

Help   Solve   Cl_ose

Fig. 15 Excel's solver tool

Exercise 19 Using Excel's Solver Package (or another appropriate solver), identify
the three equilibrium points for Eq. (24).
*Hint: To identify all three equilibrium points using Excel's solver, you must select
a value for A1 that is fairly close to the equilibrium point you wish to find. Use
Figure 13 to identify a good guess for A1 for each equilibrium point.*

Exercise 20 Perform a phase line analysis for each equilibrium point found in
Exercise 19 and classify each point as a source, sink, or node. Explain the physical
relevancy of the results ,being sure to describe the kind of climate each equilibrium
point dictates. What would it have been like to live on Earth in each case?

Finally, we consider a possible improvement of our model by incorporating data
collected by satellites about the energy radiated out by the Earth. While we

have been assuming that Earth radiates like a black body (so that the outgoing radiation follows the Stefan-Boltzmann law), Mikhail Budyko and William Sellers [8] proposed a different expression for the outgoing radiation. They proposed the following linear model for outgoing energy:

$$E_{out}(T) = A + BT, \tag{25}$$

where A and B are constants. North and Coakley [7] were able to validate this model using observational data; in particular, it was estimated that $A = 203.3\ \text{Wm}^{-2}$ and $B = 2.09\ \text{Wm}^{-2}\text{deg}^{-1}$. Here, temperatures are measured in Celsius. Thus, when we incorporate this result into our temperature evolution equation, we use $T - 273.15$ instead of just T. (Why?)

Exercise 21 Repeat Exercises 19 and 20 with the new temperature evolution model:

$$C\frac{dT}{dt} = \frac{1}{4}(1 - \alpha(T))S - (A + B(T - 273.15)), \tag{26}$$

where $\alpha(T)$ is given in Eq. (23) and A and B are given above. Compare your results with those obtained in Exercises 19 and 20.

Challenge Problem 4 One way of directly comparing Eq. (24) with Eq. (26) is by computing a linear expansion of $E_{out}^*(T) = \sigma(273.15 + T)^4$ about $T = 0$ (accounting for the fact that A and B are obtained with temperatures measured in Celsius). Compute the linearization of $E_{out}^*(T)$ and determine constants A^* and B^* so that $E_{out}^*(T) \approx A^* + B^*T$. Compare your results with Eq. (25).

Starting with the simple observation that the global average temperature at the Earth's surface increases if the amount of energy reaching the Earth exceeds the amount of energy emitted by the Earth and released into the stratosphere (and vice versa), we were able to develop models which predict the global mean temperature of Earth. More than that, we observed three different equilibrium states for the global mean temperature. How does this relate to Earth's current climate? One of the states corresponds to the current climate, while another equilibrium state was found to be unstable. The third stable equilibrium state corresponds to a deep-freeze climate, where the Earth would have been completely covered with snow and ice. In fact, this equilibrium state corresponds to a complete glaciation of the Earth, with all oceans frozen to a depth of several kilometers and almost the entire planet is covered in ice. This dramatically different Earth, for which no life could have existed, is sometimes referred to as *Snowball Earth*. There is some debate about whether Earth was completely covered with snow and ice or if there were still some "slushy" spots that could have allowed for some organisms to survive (see [6]).

5 Empirical Models for Tropical Storm Windspeeds After Landfall

5.1 Hurricane Forecasting Models

Due to the potential tragic nature of tropical systems, there is a need for the scientific understanding and modeling of these complicated phenomena in order to reduce unwanted destruction and prevent unnecessary deaths. Hurricanes are large, swirling storms with winds of 119 km per hour (74 mph) or higher and are usually characterized by a low-pressure center, a closed low-level atmospheric circulation, strong winds, and a spiral arrangement of thunderstorms that produce heavy rain. Coastal damage may be caused by strong winds and rain, high waves (due to winds), storm surges (due to severe pressure changes), and the potential of spawning tornadoes. Tropical storms also draw in air from a large area—which can be a vast area for the most severe storms—and concentrate the precipitation of the water content in that air (made up from atmospheric moisture and moisture evaporated from water) into a much smaller area. This continual replacement of moisture-bearing air by new moisture-bearing air after its moisture has fallen as rain, may cause extremely heavy rain and river flooding up to 25 miles inland from the coastline, far beyond the amount of water that the local atmosphere holds at any one time (see [1]).

The National Hurricane Center (NHC) uses forecast models (any objective tool used to generate a prediction of a future event) in order to predict and prepare for tropical storms. Specifically, they are used as a way of guiding the NHC in their preparation of official storm track and intensity forecasts. The forecast models can take on a variety of different forms of varying complexity which seek to provide predictions for the storm's track, intensity, and/or wind radii. These methods use a variety of mathematical (and statistical) techniques such as multiple regression, dynamical systems and logistic growth. Many of these models focus on one particular aspect of the storm. For instance, the LGEM, or *logistic growth equation model* is a statistical intensity model, that uses ocean heat content to predict the intensity of a tropical storm (see [1]).

In an effort to better understand and predict the path and intensity of a land-falling tropical system, we propose the development of a model for predicting the maximum sustained wind speed of land-falling tropical cyclones. The goal is to come up with a mathematical model for predicting the maximum sustained wind speed (MSWS) of a land-falling tropical cyclone. To do so, we must consider some simplifying assumptions.

Exercise 22 What are some assumptions that we can make which will aid us in developing the model? Below are some considerations. Which assumptions/considerations are most important to the development of the model?

- Sea temperatures
- Dry air

- Vertical wind shear (i.e., variation in wind velocity occurring along a direction at right angles to the wind's direction and tending to exert a turning force)
- Size of storm
- Location of storm

To build the simplest model possible, we only assume that upon making landfall, tropical cyclone winds decay rapidly and in proportion to the current strength of the cyclone. See [1] for more information. Let $v(t)$ (in miles per hour) denote the MSWS at time t (in hours) after the hurricane has made landfall.

Exercise 23 With the assumption given above, develop a differential equation which models this decay. Let α denote the proportionality constant.

Exercise 24 Solve the differential equation in Exercise 23. Let v_0 denote the initial MSWS when the tropical cyclone makes landfall (corresponding to $t = 0$). Compute $\lim_{t\to\infty} v(t)$. What is observed about the limit? Realistically, as time progresses, it has been observed that the MSWS for a tropical cyclone decreases to a nonzero limiting wind speed. Equipped with this knowledge, adjust the model in Exercise 23 so that $\lim_{t\to\infty} v(t) = v_b$, where v_b denotes this limiting nonzero wind speed.

The solution that you have obtained in Exercise 24 is a two-parameter solution in the sense that α (the decay constant) and v_b, the limiting wind speed, must be estimated from data. To this extent, consider the data given in Table 6 obtained from the National Hurricane Center, [2], for Hurricane Irma, an extremely powerful and catastrophic Cape Verde hurricane, which made landfall in the continental United States over Florida in 2017. See Fig. 16 for Hurricane Irma's track.

Table 6 Hurricane Irma's MSWS after making landfall in Florida in 2017 [2]

Time after landfall(in hours)	MSWS (in mph)	Time after landfall (in hours)	MSWS (in mph)
0	130	12	100
1	130	13	100
2	130	14	100
3	120	15	85
4	120	18	75
5	115	21	70
6	110	24	65
7	110	27	60
8	110	30	50
9	105	33	45
10	105	36	35
11	105	39	25

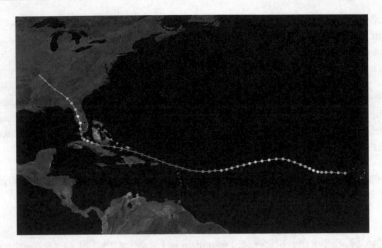

Fig. 16 The track of Hurricane Irma in 2017 [5]

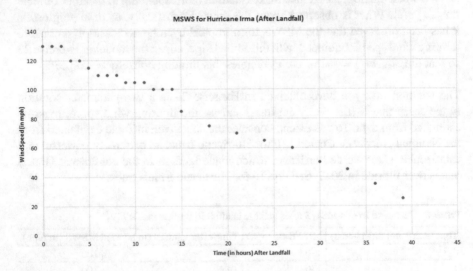

Fig. 17 Scatter plot for the wind speeds of Hurricane Irma after landfall

We wish to estimate α and v_b from this data. We plot the data in Fig. 17 of the wind speeds at different times after landfall.

We will use the Solver package in Excel to find optimal values for v_b and α that best predicts the MSWS of Hurricane Irma. To use Solver in Excel for nonlinear curve fitting, begin by creating a spreadsheet similar to the one in Fig. 18, noticing that the first two columns of the spreadsheet corresponds to the data given in Table 6. The first column of the spreadsheet contains the data corresponding to the time after landfall and the second column of the spreadsheet corresponds to the maximum sustained wind speed (MSWS) at that time.

	A	B	C	D	E	F	G	H
1	speed in mph	Time (in hours)	predicted speed in mph	chi squared		A	1	
2	130	0	3	16129		B	2	
3	130	1	3.060909068	16113.53281		C	0.03	
4	130	2	3.123673093	16097.60233		X^2	16097.6	
5	120	3	3.188348567	13644.96191				
6	120	4	3.254993703	13629.3965				
7	115	5	3.323668485	12471.60302				
8	110	6	3.394434726	11364.74655				
9	110	7	3.46735612	11349.20421				
10	110	8	3.542498301	11333.19967				
11	105	9	3.619928901	10277.91882				
12	105	10	3.699717615	10261.74721				
13	105	11	3.781936257	10245.09643				
14	100	12	3.866658829	9241.619285				
15	100	13	3.953961588	9224.841495				
16	100	14	4.043923111	9207.568692				
17	85	15	4.136624371	6538.885518				
18	75	18	4.432013724	4979.840687				
19	70	21	4.755221159	4256.881166				
20	65	24	5.108866421	3586.947881				
21	60	27	5.495815973	2970.706076				
22	50	30	5.919206222	1943.11638				
23	45	33	6.382468945	1491.313705				
24	35	36	6.889359102	790.2081317				
25	25	39	7.443985277	308.213653				
26								

Fig. 18 Spreadsheet format for using excel's solver

The solution to the differential equation obtained in Exercise 24 should have the form

$$v(t) = A + B \cdot e^{-Ct}, \tag{27}$$

where A, B, and C should depend on v_0, v_b, and α.

Exercise 25 Using *only* the data given in Table 6 as a guide, try to determine values for v_0, v_b, and α so that the computed wind speeds match the data. What factors were used to decide how to find these *optimal* values?

In Fig. 18, column C is the predicted MSWS based upon the initial guesses of v_0, v_b, and α. The formula entered into cell C2 will depend on your solution obtained in Exercise 24. For instance, if the computed solution is

$$v(t) = C + A(1 - e^{Bt}),$$

we would enter the corresponding formula:

$$= \$G\$3 + \$G\$1*(1-EXP(\$G\$2*A1))$$

and copy into all of column C, beginning with cell C2. Column D is the square of the difference between the recorded MSWS (column B) and the predicted MSWS (column C), called *chi squared* (χ^2). The following formula should be entered into cell D2 := (B2-C2) $\wedge$ 2 and copied into all of column D. Cell G4 is the sum of the chi squares values, i.e., $\sum_{i=1}^{n} \chi_i^2$. In Excel, we can write this sum as G4 = SUM(D2:D25)).

One particular way of obtaining these optimal parameter values is to *minimize* $\sum_{i=1}^{n} \chi_i^2$ since if the predicted values for the MSWS of Irma are very close to the experimental curve then the value for $\sum_{i=1}^{n} \chi_i^2$ will be small. Squared values are chosen to avoid the cancelation effect since the difference between the computed wind speed and actual wind speed could have either sign. To minimize $\sum_{i=1}^{n} \chi_i^2$, we use the "Solver" add-in in Excel to find the values of v_b and α that result in the minimum value for $\sum_{i=1}^{n} \chi_i^2$. Note that v_0 does not have to be estimated. (Why?) The procedure for using Excel's Solver is detailed in page 27.

Exercise 26 Using Excel's Solver package (or a similar computational tool), determine the *optimal* values for v_0, v_b, and α so that the computed wind speeds match the data.

If you plotted the solution to the differential equation obtained in Exercise 24 with the optimal values obtained in Exercise 26, one should get a graph that resembles Fig. 19.

Exercise 27 Using your graph as a guide, explain how your results verify the following observed claim regarding tropical storms:

Tropical storms whose circulations are partially over water (close to the coastline) decay less rapidly than those that are entirely over land.

Why do you think this is the case? How could you adjust your model to account for this observation?

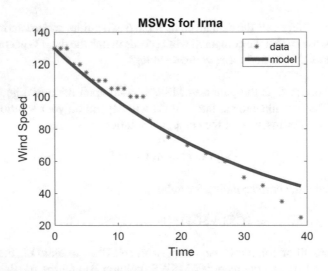

Fig. 19 Predicted MSWS vs. Actual MSWS for Irma

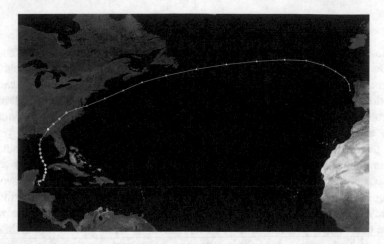

Fig. 20 The track of Hurricane Michael in 2018 [4]

Table 7 Michael's MSWS after making landfall in Florida in 2018 [3]

Time after landfall (in hours)	MSWS (in mph)	Time after landfall (in hours)	MSWS (in mph)
0	160	10.5	70
0.5	155	13.5	60
1.5	150	16.5	50
2.5	140	19.5	50
3.5	125	22.5	50
4.5	115	25.5	50
5.5	100	28.5	50
6.5	90	31.5	50
7.5	85	34.5	50
8.5	80	37.5	60
9.5	75	40.5	65

Challenge Problem 5 During the 2018 hurricane season, Hurricane Michael roared ashore near the Florida Panhandle with a maximum sustained wind speed of 155 mph. Hurricane Michael formed in the Gulf of Mexico and made its way northwards until it made landfall near Panama City, FL. This made Hurricane Michael one of the strongest storms to ever hit the United States–and the first Category 5 storm to hit the United States since Hurricane Andrew in 1992. See Fig. 20 for Hurricane Michael's track. Consider the data in Table 7, obtained from [3], which gives the MSWS for Hurricane Michael:

Using the previous problems as a guide, develop a mathematical model which determines the MSWS for Hurricane Michael as a function of time. Construct a plot similar to Fig. 19. Discuss your results. Is there an issue of "overpredicting" (predicted wind speeds higher than the actual wind speed) and/or "underpredicting" (as observed with the Irma model)? What possible adjustments could be made to this model to increase its accuracy?

6   Suggested Projects

Now that we have considered several examples related to climate and weather, you are now in a position to attempt some research level projects using the tools (both mathematical and numerical) developed in this chapter. Because several examples were driven by data, many projects can be started by accessing some interesting data collected on behalf of understanding the behavior of climate and/or weather. In any case, the best starting points are to either graph the data points (if you are using data to guide your research) or to make suitable assumptions about your problem that will aid in the development of a model. Many of the projects will be some combination of modeling and data analysis. For instance, one can do work on developing a mathematical model (or adjusting a known model), running different simulations of the model, investigating equilibrium solutions, and tuning the model. On the other hand, you may find yourself analyzing some type of interesting data by performing a regression analysis in an attempt to draw some conclusion from the data and make some suitable prediction.

Research Project 1 The South Pole Observatory is situated at the geographic South Pole and has been collecting data since 1957. It is one of the four major observatories operated by the NOAA Earth System Research Laboratory, Global Monitoring Division (GMD). Ozone data collection began in 1963 and CO_2 concentrations started being measured in 1975. Using the data provided at this link (*C02Data-SPO.xlsx*):

https://1drv.ms/x/s!Aq7IzuMtCFE8nVjn6LWTnuniNOn0?e=QhefAK

perform a linear and quadratic regression analysis. Compare and contrast your findings from the Mauna Loa Observatory.

Research Project 2 Model the intensification of tropical cyclone wind speeds during the period of time where a tropical cyclone develops over warm waters.

Research Project 3 Study the effects of human activity on the climate by focusing on how carbon emissions are effected by deforestation, land use, agricultural practices, population growth and globalization.

Research Project 4 Model the climate using an energy balance model of a planet other than Earth.

Research Project 5 Model some specific physical process of the climate system, potentially studying how it contributes to the overall climate (e.g., Arctic sea ice, ocean circulation, biosphere, or permafrost).

References

1. Emanuel, K.;*Divine Wind: The History and Science of Hurricanes* . Oxford University Press (2005)
2. Maximum Sustained Wind Speeds for Hurricane Irma. *Weather Underground* TWC Product and Technology, https://www.wunderground.com/hurricane/atlantic/2017/hurricane-irma?map=history. Accessed 31 March 2019
3. Maximum Sustained Wind Speeds for Hurricane Michael. *Weather Underground* TWC Product and Technology, https://www.wunderground.com/hurricane/atlantic/2018/Post-Tropical-Cyclone-Michael. Accessed 18 July 2019
4. Beven II, J., Berg,R., and Hagan, A.; "Tropical Cyclone Report: Hurricane Michael." *National Hurricane Center* NOAA, https://www.nhc.noaa.gov/data/tcr/AL142018_Michael. pdf Accessed 18 July 2019.
5. Cangialosi, J., Latto,A., Berg, J.; "Hurricane Irma (AL112017) Tropical Cyclone Report." *National Hurricane Center* NOAA, https://www.nhc.noaa.gov/data/tcr/AL112017Irma.pdf Accessed 18 July 2019.
6. Engler, H., Kaper, H.;*Mathematics and Climate* . Society for Industrial and Applied Mathematics (2013)
7. North, G., Coakley, J.; *Differences between seasonal and mean annual energy balance model calculations of climate and climate sensitivity*, Journal of the Atmospheric Sciences, 36 (1979), pp. 1189-1204.
8. Budyko, M. I.; *The effect of solar radiation variations on the climate of the Earth*, Tellus, 21 (1969), pp. 611-619
9. "The Greenhouse Effect". *University Corporation for Atmospheric Research* https://scied.ucar.edu/longcontent/greenhouse-effect Accessed 21 July 2019.
10. Sellers, W.D.; *A Global Climatic Model Based on the Energy Balance of the Earth-Atmosphere System*. Journal of Applied Meteorology (1969), no. 8, 392–400.
11. North, G.T.; *Theory of Energy-Balance Climate Models*. Journal of the Atmospheric Sciences (1975), no. 32, 2033–2043.
12. North, G.T.; *Analytical Solution to a Simple Climate Model with Diffusive Heat Transport*. Journal of the Atmospheric Sciences (1975), no. 32, 1301–1307.
13. Manabe, S., Strickler, R.F.; *Thermal Equilibrium of the Atmosphere with a Convective Adjustment*, Journal of the Atmospheric Sciences (1964) no. 21, 361–385.
14. Manabe, S., Wetherald, R.T.; *Thermal Equilibrium of the Atmosphere with a Given Distribution of Relative Humidity*, Journal of the Atmospheric Sciences (1967), no. 24, 241–259.
15. Kasting, J.F., Whitmore, D.P., Reynolds, R.T.; *Habitable Zones around Main Sequence Stars.*, Icarus (1993), no. 101, 108–128.
16. Pierrehumbert, R.T.; *The hydrologic cycle in deep-time climate problems*, Nature (2002), no. 419. 191–198

17. Stone, P.H. *Constraints on dynamical transports of energy on a spherical planet.* Dynamics of Atmospheres and Oceans (1978), no. 2, 123–139.
18. Emauel, K. *A simple model of multiple climate regimes.* J. Geophys. Res. (2002), no. 107, 10 PP.
19. Pales, J., Keeling, C.D. *The Concentration of Atmospheric Carbon Dioxide in Hawaii* Journal of Geophysical Research (1965), no. 70, 6052–6076.
20. Le Quere, C. , Raupach, M.P., Canadell, J.G., Marland, G., et al. *Trends in the sources and sinks of carbon dioxide* Nature Geoscience, no. 2, 831–836.

Beyond Trends and Patterns: *Importance of the Reproduction Number from Narratives to the Dynamics of Mathematical Models*

Aditi Ghosh and Anuj Mubayi

Abstract

How society talks about important current affairs such as spread of black lives matter ideology, interventions to curb surge in COVID-19 infections, and breakthrough in global ambition on climate has serious consequences. The centrality of narrative in the pursuit of resolving medical, social, and life sciences challenges has a common theme and is often shaped by a simple term referred to as tipping point or reproduction number. The reproduction number, a key parameter in understanding long-term implications of a problem, is defined as a number indicating a point beyond which there are significant changes in system's behavior. For example, in the case of the spread of an infectious disease, the reproduction number characterizes the transmission potential of an epidemic growth, beyond which either it increases exponentially or diminishes to a negligible impact. Hence, it is often interpreted and used to inform the potential effectiveness of intervention strategies. In this chapter, we focus on applications in epidemiology and public health and provide a variety of reproduction numbers and their interpretation. We also discuss the important concept of testing and dispersion number and their practical relevance during an epidemic.

A. Ghosh (✉)
Department of Mathematics, Texas A&M University, Commerce, TX, USA
e-mail: Aditi.Ghosh@tamuc.edu

A. Mubayi
PRECISIONheor, Los Angeles, CA, USA

Center for Collaborative Studies in Mathematical Biology, Illinois State University, Normal, IL, USA

Intercollegiate Biomathematics Alliance (IBA), Normal, IL, USA

© The Author(s), under exclusive license to Springer Nature Switzerland AG 2022
E. E. Goldwyn et al. (eds.), *Mathematics Research for the Beginning Student,*
Volume 2, Foundations for Undergraduate Research in Mathematics,
https://doi.org/10.1007/978-3-031-08564-2_9

Suggested Prerequisites *Knowledge of basic notions of ordinary differential equations, numerical simulations, and linear algebra. Some acquaintance with public health, mathematical epidemiology, as well as experience with mathematical modeling and simulations will be helpful but are not required.*

1 Introduction

The critical value of the reproduction number also commonly referred to as the tipping point is a naturally occurring issue, which can be easily seen in many applications. Using examples from everyday life, Malcolm Gladwell [9] attempts to explain it in his 2000 book *A Tipping Point: How Little Things Can Make a Big Difference*. However, it has long history of mathematical definition, generally based on threshold condition, which was rigorously quantified and explained using mathematical methods for epidemics by Kermark and Meckdrink in their seminal paper in 1927 [15]. Kermack and McKendrick described an infection spread in population and capturing the number of people infected with a contagious illness in a closed population over time. They explained the rapid rise and fall in the number of infected patients observed in epidemics such as the plague (Bombay 1906) and cholera (London 1865) using a quantity, referred later as the reproduction number. On the other hand, social science journalist Gladwell collected a set of common, day to day, examples to define a similar process via a tipping point and interpreted this as "the moment of critical mass or the boiling point." His book tries to explain and describe the sudden sociological changes in many real-life scenarios. For example, some products and ideas quickly turn into widely accepted notions and spread like wildfire, while others get doomed.

Smallpox Eradication from the World Through Vaccination Smallpox killed millions of individuals during the twentieth century and was eventually eradicated in 1980 primarily through mass vaccination efforts. Although the initial steps toward vaccines took place in 1796 when Edward Jenner injected infected puss in his gardener's eight-year-old son. It was not until the twentieth century when a threshold in vaccinating critical size of population was achieved for smallpox that resulted in natural disappearance of the disease from the population. It was essentially a combination of many interventions including active surveillance and case finding, contact tracing, ring vaccination (controlling an outbreak by vaccinating a ring of people around each infected individual), and communication campaigns to find, track, and inform affected people, which played a role in eradication of smallpox. For more details on a smallpox vaccine model, please refer to [4]

Arab Spring in Middle Eastern Countries A similar example was noted during the well-known Arab spring protest in 2010. In general, peaceful protests are regularly seen in democratic countries, as a way to show unhappiness of locals to government decisions impacting them. Often, these protests die down, either because of not enough response of population for the movement or due to decision makers accepting the demands of protesters. One such protest leads to the series of protests and demonstrations across the Middle East and North Africa that commenced in 2010 and became known as the "Arab Spring." It was sparked by the first protests that occurred in Tunisia on 18 December 2010 following a young Tunisian's self-immolation in the protest of police corruption and ill treatment. This was a tipping point for a big movement ahead. With success of this protest above a critical population size, a wave of civil uprisings erupted in many countries in the Middle East such as Algeria, Jordan, Egypt, Yemen, Libya, Bahrain, Syria, Iraq, Kuwait, Morocco, Oman, Sudan, etc. These unrest resulted in some rulers been forced out of power by the end of February 2012. See [7] for more details on developing a riot model based on Arab Springs.

Chicago Crime in 1990s and Midnight Basketball Games in the USA A similar example served as a tipping point when a social worker named Van Standifer suggested a program referred to as midnight basketball games, an initiative in the 1990s for opening city parks with lights from 10pm to 2am (known peak crime hours), to curb inner-city crime in the United States by keeping urban youth off the streets and engaging them with alternatives to drugs and crimes. For few years, the operation was small, but then with help from inspiring community leaders the story of midnight basketball leagues spread like a wildfire and quickly the numbers swelled above the threshold level. It became a norm, and drastic reduction in crime rates in many parts of the USA was observed. An offbeat idea served as a tipping point to bring down the crime rates in the Chicago area. [21] provides a detail understanding of the spatial dynamics of neighborhood transition in the Chicago crime case.

Tobacco Smoking Trends in the USA In another example related to cigarette smoking patterns, there has been a significant decline in cigarette smoking in the USA (according to CDC, 13.7% of adults smoked cigarettes in 2018, as compared to 20% in 2005), a trend never seen before. This has been possible due to consistent and coordinated effort by the public health community to ban smoking in public places and inside buildings. Initially, these bans were ineffective as they were not consistent throughout the country. However, later countrywide bans (above a critical number of bans in local regions) accounted for a decline in smoking. It made smoking seem like a more stigmatizing behavior and also reduced secondhand smoke incidents in closed confined places and/or left a smoker with a high social cost of leaving social contexts to smoke at the expense of continued participation in social gathering.

Symbiotic Relationship Between Rice and Rats in Southeast Asia Rodents species, a key mammals in human life for centuries, have had impacts on many crops like rice in Southeast Asia. This has led to the development of many ecological based rodent management (EBRM) programs, effective community actions to control population of rice field rats. The programs are designed based on the understanding of the breeding dynamics of the rice field rat and its relation to the rice planting season. Rodents play an important role in food chains. For example, they are critical ingredients to engineer ecosystems, such as their boroughs and trails influence the flow of waters and nutrients at local scales, their collections and hoarding of seeds in forests can lead to generation of new trees away from competing trees, and they regulate the population growth of some weeds and insects as they are part of their diet. Therefore, we need to protect certain species to have healthy ecosystem. However, rodents are often pests and can cause severe damage to rice crops for small farmers. The rice field rats breed according to rice seasons. Hence, under EBRM program, rice rats are controlled before the breeding season begins and when critical threshold density in rat population is reached. This threshold is the demographic reproduction of the rat population beyond which control is necessary to save rice fields from damage by rat pests. This article [30] provides in-depth insight to the rice and rat problem.

These examples show that the tipping point or reproduction number is a common phenomenon that is regularly observed around us in many forms. There is a need to systematically describe and understand this phenomenon, so that future patterns of a problem can be predicted and controlled. The success of any kind of social, ecological, or public health epidemic is heavily dependent on the involvement of people with a particular and rare set of social gifts, which are often driven by the "80/20 Principle" (that is, roughly 80% of the "work" will be done by 20% of the participants; Pareto Principle of Economics). Mathematically, the 80/20 rule roughly follows a Pareto distribution (see[2]) for a particular set of parameters, and many natural phenomena have been shown empirically to exhibit such a distribution. The distribution can be written as

$$P(X > x) = \begin{cases} (\frac{K}{x})^{\alpha} & x \geq K \\ 1 & x < K, \end{cases}$$

where K (also called as scale parameter) is the positive minimum possible value of X, and α (also referred to as shape) is a positive parameter.

Gladwell explains the tipping point phenomenon using three rules that are critical for the existence of an outbreak: (1) The Law of the Few, (2) The Stickiness Factor, and (3) The Power of Context. The Law of the Few states that a certain small set of people can be considered as Connectors, who know large numbers of people and who are in the habit of making introductions (i.e., introducing people who work or live in different community circles; for example, as seen in Milgram's experiments in the small-world problem and the Six Degrees of Kevin Bacon game). These individuals have vast and varied information and also are good in persuading others with their negation skills. The Stickiness Factor refers to the specific content of an information that renders its impact memorable. For example, the popular movie like *Contagion* pioneered the properties of the stickiness factor, thus enhancing effective retention of educational content as well as entertainment value. The Power of Context is used for knowing how sensitive individuals are to certain environments or contexts. For example, epidemics are sensitive to the environmental conditions

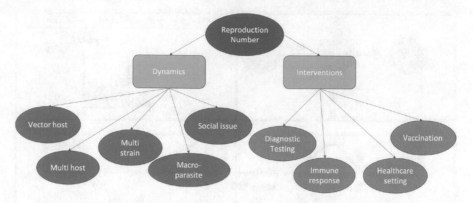

Fig. 1 Various applications of the reproduction number in this chapter

and circumstances of the times and places in which they occur. There is a social psychological theory, called as the bystander effect, or bystander apathy, which states that individuals are less likely to offer help to a victim when there are other people present. This effect has been observed in many settings such as involvement of subordinates and managers in a workplace environment.

In this chapter, we have tried to collect examples from real life and provided conceptualization of the reproduction number or tipping point quantity. The remaining chapter is structured as given in Fig. 1.

2   Concept of Reproduction Number for an Epidemic

Using basic concepts from calculus, we can write equations that can capture the epidemic of a disease. The basic model compartmentalize a given population into three groups: Susceptible (those who can catch the disease; S), Infectives (those who can spread the disease; I), and Recovered (those who are immune and cannot spread the disease; R). Hence, it is often referred to as the SIR model (see the flow chart of the SIR model in Fig. 2). We then describe certain modeling assumptions for formulating our model. The assumptions in the SIR model are:

- The population is large, fixed in size, and confined to a well-defined region.
- The population is well mixed. That is, everyone is equally likely to come in contact with the same fraction of people in each category every day.
- The compartments are represented by state variables (for example, S, I, and R are variables that change over time, t) and transition rates between the compartments. The transition rates consist of parameters (which are constant over time) and state variables.
- The rate of change of number of infected individuals is equal to the sum of rate of transmission of infection and rate of recovery. That is,

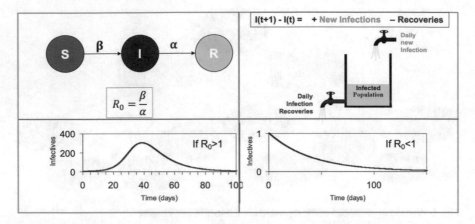

Fig. 2 Interpretation of SIR epidemic model. The four frames are explained as follows. *Top left:* the flowchart representing the flow between different sub-populations ($S- >$ Susceptibles; $I- >$ Infectious; $R- >$ Recovered) of the epidemic model, referred to as compartments, along with the corresponding reproduction number formula. *Top right:* caricature of influx and outflux in a representative compartment; the bucket represents each of the compartments in the model. *Bottom left* and *Bottom right*: temporal dynamics of the disease when $R_0 < 1$ and $R_0 > 1$

$$\text{Rate of change in infectives} = \text{Transmission rate} - \text{Recovery rate} \qquad (1)$$

$$\frac{dI(t)}{dt} = \beta S(t)\frac{I(t)}{N} - \alpha I(t) \qquad (2)$$

- New infections cause the size of S to decrease and the size of I to increase by the same amount. Recovery results in decrease in the size of I and increase in the size of R by the same amount. The number of diseases causing contacts per day is given by $\beta SI/N$, where only the contacts of infectives with susceptible people are considered and this process is captured by the product of the size of S and the size of I.

The basic reproduction number, R_0, is a threshold parameter for disease extinction or survival in isolated populations. The basic reproduction number, R_0, is defined as the expected number of secondary cases (infection is transmitted to new people) produced by a single (typical) infection in a completely susceptible population. In other words, in the case of no disease transmission from other source, the basic reproduction number, R_0, is the average number of secondary cases arising from each primary case in a susceptible population. The importance of basic reproduction number is that it determines whether a disease will fade out or spread through person-to-person transmission. Implementation of non-pharmaceutical or medical interventions can control transmission by reducing the basic reproduction number below 1, thus causing the disease to fade out. The basic reproduction number can be defined as

$$R_0 = \tau c d$$

where τ is the transmissibility (i.e., probability of infection given contact between a susceptible and infected individual), c is the average rate of contact between susceptible and infected individuals, and d is the duration of infectiousness [13, 27]. Note that, in the SIR model, $\beta = \tau c$ and $d = \frac{1}{\alpha}$.

To understand this analytically, consider an epidemic model—SIR model, where SIR stands for Susceptible–Infected–Removed (Fig. 3a). We assume a homogeneous population where any infected individual has a probability of contacting any susceptible individual and a closed population of N individuals with no birth and death and say that S are susceptible, I are infected, and R are removed. Write $s = \frac{S}{N}, i = \frac{I}{N}, r = \frac{R}{N}$ to denote the fraction of the total population, N, in each compartment. We consider β as the effective contact rate and α the removal rate and assume all rates are constant. The SIR model is then

$$\frac{ds}{dt} = -\beta s i$$

$$\frac{di}{dt} = \beta s i - \alpha i$$

$$\frac{dr}{dt} = \alpha i \tag{3}$$

When does an outbreak or epidemic occur? An epidemic occurs if the number of infected individuals increases, i.e., $\frac{di}{dt}$ is positive which implies $\beta s i - \alpha i > 0$. For an epidemic onset, nearly everyone (except the index case) is susceptible. Hence, $s = 1$, substituting where, we have $R_0 = \frac{\beta}{\alpha} > 1$.

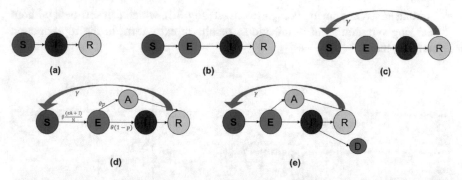

Fig. 3 Assumptions and structure of some models: (**a**) model with lifelong immunity (e.g., Chicken pox), (**b**) model with incubation period (e.g., Ebola), (**c**) model with temporary immunity (e.g., Influenza), (**d**) model with pre-symptomatic and asymptomatic stages (e.g., Tuberculosis), and (**e**) model with infection related deaths (e.g., COVID-19)

2.1   Characteristics and Dynamics of an Epidemic Model

The SIR epidemic type frameworks are simple but useful models. One can do sufficiently enough analytical exploration with these epidemic models while ensuring that they capture salient features of epidemiological and biological processes. The assumptions of a model describe its flow/rates, structure, and mathematical representation. For example,

- For some diseases, incubation or latent stage (represented by adding an extra compartment E, Exposed, in the SIR model; Fig. 3b) is critical and hence, parameters related to incubation period (i.e., duration that an initial infected case remain noninfectious) becomes relevant to containing an epidemic. The resulting SEIR model has dynamics similar to that of SIR model except the epidemic is delayed with lower peak in the SEIR model as compared to the SIR model (Fig. 4a). Adding E class does not change the total number of people who become infected, but it significantly delays and widens the peak of infectious cases, thus flattening the curve.

- In some diseases, infected individuals recover but only remain temporary immune to infection after recovery (represented by parameter γ; Fig. 3b). There are two major features observed by adding temporary immunity component: (1) we replenish the susceptible compartment that can become infected again, and hence, infection becomes endemic (i.e., outbreak never dies out) and (2) there is possibility of having several peaks before the steady state is reached, which depends on the value of rate of losing immunity, γ, is sufficiently small (immunity is longer lasting; Fig. 4c). The appearance of the peak is due to longer time by an individual in the recovered compartment, which allow the epidemic to follow most of its original trend before a number of susceptibles start to grow again.

- Adding an asymptomatic compartment (A; Fig. 3d), which can self-heal without showing symptoms but is infectious, results in extra term in the reproduction number, that is,

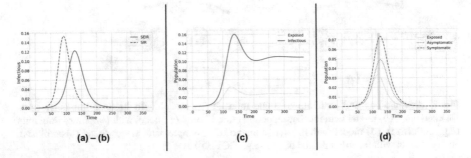

Fig. 4 Dynamics of some models: (**a**)–(**b**) comparison of SIR and SEIR model, (**c**) SEIR model with temporary immunity, and (**d**) SEIRS model with asymptomatic compartment

$$R_0 = \frac{\beta}{\alpha}(\epsilon p + (1 - p))$$

where α is the per capita recovery rate of A or I individual. Depending on the difference between the recovery rates of $A \rightarrow R$ and $I \rightarrow R$, the peaks in A and I compartments, respectively, would occur at different times (Fig. 4d assumes the same recovery rate). Moreover, it also shows that as we add more and more compartments to our models, the smaller the population of each individual compartment becomes.

- When disease death compartment is added (i.e., class D is added), the number of Recovered individuals (i.e., size of R compartment) naturally reduces as compared to the trends observed in the model without disease deaths.

As the epidemic progresses, more and more information is gathered, and more detailed models can be used. This constant refinement also helps improve the reliability of the scenarios and the decision processes. In the case of Fig. 3b where E is the fraction of exposed individuals (infected but are not yet infectious) and R the recovered individuals being permanently immune gives us the following model:

$$\frac{dS}{dt} = \beta SI + \lambda - \mu S$$

$$\frac{dE}{dt} = \beta SI - (\mu + k)E$$

$$\frac{dI}{dt} = kE - (\gamma + \mu)I$$

$$\frac{dR}{dt} = \gamma I - \mu R \tag{4}$$

where β is the effective contact rate, λ is the birth rate of susceptibles, μ is the mortality rate, k is the progression rate from exposed (latent) to infected, and γ is the removal rate.

Exercise 1 Can you construct a model for Fig. 3d?

2.2   Epidemiological Interpretation of R_0

As we discussed earlier, the basic reproduction number, R_0, is defined as the average number of secondary infections produced by a typical case of an infection in a population where everyone is susceptible. The basic reproduction number is calculated in the absence of interventions during an outbreak. The effective reproduction number is calculated when the susceptible population is not the same percent. Thus, an important concept, the effective reproduction number denoted by R_t, is the expected number of new infections caused by an infectious individual in a population where some individuals may no longer be susceptible. It is used for assessment of changes in health policy, population immunity, and other factors that have impacted transmission at specific places in time. R_t is given by the formula $R_t = s(t)R_0$, where R_0 is the basic reproduction number and $s(t)$ is the proportion of susceptible individuals in the population at time "t". We have that $R_t \leq R_0$, with the upper bound the basic reproduction number only being achieved when the entire population is susceptible. Similarly, the quantity, controlled reproduction number, R_c, describes the ability of disease spreading when interventions (such as quarantine, isolation, or vaccine) are taking place from the starting of the outbreak. It is a measure that reflects the impact of an intervention in order to reduce R_c. Note that the disease will decline and eventually die out if $R_c \leq 1$.

Now, let us understand what do we mean by generations. In epidemic models, generations are the waves of secondary infection which flow from each previous infection. So, the first generation of an epidemic is the number of secondary infections resulting from infectious contact with the index case, who is of generation zero. If R_i denotes the reproduction number of the ith generation, then R_0 is simply the number of infections generated by the index case, i.e., generation zero. In complex models, it is not easy to find reproduction numbers intuitively, and hence we use a method called *next generation matrix method* [12, 25, 34].

2.3 Computation of Reproduction Number Using Next Generation Method

We consider a system of differential equations describing an epidemic process and follow the following steps to compute the reproduction number (see the details in [33]):

1. First, separate its state variables (e.g., S, I, and R in SIR epidemic model system) into two groups: (1) the infected group (e.g., I) and (2) the noninfected group (e.g., S and R). The equations are then reordered to include infected group first followed by noninfected group.
2. Then, the system is rewritten in vector-matrix form as follows:

$$x'(t) = f(x(t))$$

where $x(t) = (x_1, x_2, \cdots, x_n)^T$ and function $f : R^n \rightarrow R^n$ is partitioned as below:

$$f(x) = \mathscr{F}(x) - \mathscr{V}(x)$$

where $\mathscr{F}$ and $\mathscr{V}$ are the non-negative vectors. The component of vector, $\mathscr{F}$, is $\mathscr{F}_i$ which consists of the rate of **newly** infected individuals in the compartment i.

3. We then compute *disease-free equilibrium* (DFE) of the system, which occurs when the number of infected individuals in the population is zero. Let DFE be given by $x^* = (x_1^*, \ldots, x_p^*, 0, \ldots, 0)$ (e.g., the DFE of SIR model is $x^* = (S^*, R^*, 0)$, where $I^* = 0$).

4. Next, we compute two matrices as below:

$$D_x^* \mathscr{F} = \begin{bmatrix} 0 & 0 \\ 0 & F \end{bmatrix}$$

$$D_x^* \mathscr{V} = \begin{bmatrix} J_1 & J_2 \\ 0 & V \end{bmatrix}$$

where D_x^* represents partial derivatives of $\mathscr{F}$ and $\mathscr{V}$ with respect to vector of state variables, x, computed at DFE (x^*).

5. Since F is non-negative and V is a non-singular M-matrix, V^{-1} can be computed and is non-negative. Therefore, we can compute FV^{-1}, which will be also non-negative.

   Note that the (j, k) entry of V^{-1} matrix will represent the average length of time the kth individual spends in compartment j during its lifetime, assuming that the population remains near the DFE. The (i,j) entry of F is the rate at which infected individuals in compartment j produce new infections in compartment i. Hence, the (i,k) entry of the product FV^{-1} is the expected number of new infections in compartment i produced by the infected individual originally introduced into compartment k.

6. We call FV^{-1} the next generation matrix for the model and $R_0 = \rho(FV^{-1})$, where $\rho(A)$ denotes the spectral radius of a matrix A.

   Note that the spectral radius of a matrix M is defined as

$$\rho(M) = max\{|\lambda|, \lambda \text{ is an eigen value of } M\}.$$

The process of computation of reproduction number and its expression can be interpreted using DFE and its stability. The DFE, x_0, is locally asymptotically stable if all the eigenvalues of the matrix $Df(x_0)$ (i.e., Jacobian matrix of right-hand side of the system) have negative real parts and unstable if any eigenvalue of $Df(x_0)$ has a positive real part. Thus, if x_0 is a DFE of the model system, then x_0 is locally asymptotically stable if $R_0 < 1$, but unstable if $R_0 > 1$. In different cases, multiple DFEs can exist, and in that situation, it is possible to calculate an R_0 value related to each equilibrium. The definition of a global R_0 is much more complicated and requires a case-by-case study.

Exercise 2 A disease begins to spread in a village population of 800. The infective period has an average duration of 16 days and an average infective is in contact with 0.2 person per day. What is the basic reproduction number? To what level must the average rate of contact be reduced so that the disease will fade out?[1]

Exercise 3 A disease is introduced by two visitors into a town with 1500 inhabitants. An average infective is in contact with 0.3 inhabitant per day. The average duration of the infective period is 5 days, and recovered infectives are immune against reinfection. How many inhabitants would have to be immunized to avoid an epidemic?[1]

Exercise 4 Many diseases, such as HIV and tuberculosis, have latent periods. After infection, the susceptibles are first in the latent period of the disease. They cannot transmit the disease in this period. As time passes or for other reasons, the individuals in the latent period may become infectious and will transmit the diseases to susceptible people. Now, we construct an $SLIR$ model. Here, S stands for the susceptible class, L stands for the latent class, I stands for the infectious class, and R stands for the recovery (treated) class. The system of ordinary differential equation is then of the form:

$$\frac{dS}{dt} = -\beta S \frac{I}{N} + \mu N - \mu S$$

$$\frac{dL}{dt} = \beta S \frac{I}{N} - \mu L - \delta L$$

$$\frac{dI}{dt} = \delta L - \mu I - \gamma I$$

$$\frac{dR}{dt} = \gamma I - \mu R \tag{5}$$

where $N = S + L + I + R$.

(a) Reduce the system to a 3D system. (Hint: does the R class contribute to the other classes? Does N change with time?)
(b) Find the basic reproduction number.
(c) Explain the meaning of the basic reproduction number you found in (b).

Exercise 5 If vaccination strategies are incorporated for newborns, we assume that not every newborn is susceptible. Suppose that the per capita vaccination rate is p. Then, any newborn is vaccinated with probability p. The modified model is the following:

$$S'(t) = (1 - p)\mu N - \beta S \frac{I}{N} - \mu S$$

$$I'(t) = \beta S \frac{I}{N} - (\mu + \gamma)I$$

$$V'(t) = \gamma I - \mu V + p\mu N \tag{6}$$

(a) Calculate the Jacobian matrix of (6) at the disease-free equilibrium points. (Hint: no infected individuals in steady state)
(b) Find the basic reproduction numbers (R_0).

Exercise 6 Consider the model system given in Eqs. 23, and compute the virus-free steady state as well as show that the

$$N_{cri} = \frac{(\mu_V + kT_H^*)(\mu_I + \delta T_K^*)}{kT_H^*}$$

is the threshold quantity of the system, similar to the reproduction number.

Exercise 7 Consider the model system given in Eqs. 18, and use the next generation method to show that the reproduction number is

$$R_0 = \frac{\lambda \beta H}{(\mu + d + \alpha)(\gamma + \beta H)}.$$

Note that free-living parasite larvae introduced (λ) in the environment (W) need to survive its duration in the environment (survival duration is $1/(\gamma + \beta H)$) while spreading to human host (H)

---

3 Application of Reproduction Number for Understanding Dynamics

3.1 The Reproduction Number for the Single-Host-Vector Model

Suppose there is a vector-borne disease that is spread by a vector species in a host population. The simplest model for such vector-borne disease is as follows (for examples, similar models are described in studies such as [10, 13, 29, 32]):

$$X' = mabY(1 - X) - rX \tag{7}$$

$$Y' = acX(e^{-gn} - Y) - gY \tag{8}$$

Here, a is the expected number of bites on humans per vector, m is the equilibrium vector density per human (the ratio of the number of hosts to vectors per unit area), n is the length of incubation period (per unit time), g is the daily force of vector mortality (per unit time), b is the transmission efficiency from infectious vector to susceptible human, c is the probability of infection of an uninfected vector by biting an infectious human, X is the proportion of infected humans, and Y is the proportion of infectious vectors. We use the next generation matrix to calculate R_0:

$$F = \begin{bmatrix} 0 & abm \\ ace^{-gn} & 0 \end{bmatrix}, \text{ and } V = \begin{bmatrix} -r & 0 \\ 0 & -g \end{bmatrix}$$

Hence,

$$FV^{-1} = \begin{bmatrix} 0 & \frac{-abm}{g} \\ \frac{-ace^{-gn}}{r} & 0 \end{bmatrix}$$

Thus, the reproduction number of single-host-vector system is given by spectral radius of the FV^{-1}, that is,

$$R_0 = \sqrt{\frac{mab}{g} \cdot \frac{ace^{-gn}}{r}} = \sqrt{R_0^{vh} R_0^{hv}}$$

$$= \sqrt{\frac{ma^2 e^{-gn} bc}{rg}} \tag{9}$$

where the reproduction number from host to vector, R_0^{hv}, represents the number of vector infections caused by one infected host and the reproduction number from vector to host, and R_0^{vh} represents the number of host infections caused by one infected vector [36]. The square root represents the geometric mean that takes the average number of secondary host (or vector) infections produced by a single infected host (or vector).

3.2   The Reproduction Number in Multi-host Setting

Suppose there is a system capturing transmission dynamics of a vector-borne infectious disease, which is spread in two host populations via a vector population. Let each of the host populations be further divided into susceptible, infectious, and recovered groups, while the vector population is subdivided into susceptible and infectious groups. In this situation, another category of reproduction is defined to analyze the dynamics of the system. This number is called as the type-reproduction number (R_T) for the specific host type and can be interpreted as the average number of secondary cases of that type produced by the primary cases of the same host type during the entire course of infection [23]. It takes into account two types of secondary infections, cases directly transmitted from the specific host and the cases indirectly transmitted by way of other types, who were infected from the primary cases of the specific host with no intermediate cases of the target host. This number is useful to design specific control program for a particular host species. The next generation matrix of the system is

$$K = \begin{bmatrix} 0 & K_{12} & 0 \\ K_{21} & 0 & K_{23} \\ 0 & K_{32} & 0 \end{bmatrix}$$

where K_{ij} represents the average number of new infections among the susceptible of type i, generated by an infected of type j. Hence, the basic reproduction number of the disease is

$$R_0 = \rho(K) = \sqrt{K_{12}K_{21} + K_{32}K_{23}} \tag{10}$$

$$= \sqrt{R_0^{h_1 v_2} R_0^{v_2 h_1} + R_0^{h_3 v_2} R_0^{v_2 h_3}} \tag{11}$$

Here, there are two hosts (Type 1 and Type 3) and one vector (say Type 2) in the system. Note that, in a vector-borne disease, infected host of Type 1 cannot directly infect susceptible host of Type 3, or vice versa, it is infected via vector represented here by Type 2.

The type-reproduction number for Type 1 is given by

$$R_{T1} = K_{12}K_{21} + K_{12}(K_{32}K_{23})K_{21} + K_{12}(K_{32}K_{23} \cdot K_{32}K_{23})K_{21} + \ldots \tag{12}$$

$$= \frac{K_{12}K_{21}}{1 - K_{32}K_{23}} \tag{13}$$

The type-reproduction numbers of Type 2 and Type 3, respectively, are

$$R_{T2} = K_{12}K_{21} + K_{32}K_{23} \tag{14}$$

$$R_{T3} = \frac{K_{32}K_{23}}{1 - K_{12}K_{21}} \tag{15}$$

Note that if $R_0 < 1$ and R_{Ti} for all host type i is greater than 1, then host type i is a reservoir of infection.

3.3 Invasion Reproduction Number in Multi-strain Model

Suppose there is a disease with more than one strain of a pathogen cocirculating within human populations. Consider a simple competitive multi-strain extension of our basic SIS model with reservoir exposure. Each strain has its own transmission parameter (β_k) and recovery rate (γ_k). We assume that infection with one strain prevents infection from all other strains for the duration of the infection. Suppose there are n strains, then the system is given by

$$s' = -\sum_{k=1}^{n} \beta_k i_k s + \sum_{k=1}^{n} \gamma_k i_k \tag{16}$$

$$i_k' = \beta_k i_k s - \gamma_k i_k, \quad k \in \{1, 2, \ldots, n\} \tag{17}$$

The basic reproduction number in a fully susceptible population is $R_0^k = \frac{\beta_k}{\gamma_k}$. If we account for strain competition, it can lead to the extinction of strains that would otherwise persist in a population. Therefore, we consider the invasion reproduction number for each strain, $\tilde{R}^k$, i.e., not the reproduction number in a fully susceptible population, but in a population at endemic equilibrium with all the other strains. It is difficult to compute $\tilde{R}^k$ explicitly but can be obtained by setting the equilibrium values of all strains except any infections of strain k [19]. The basic reproduction number of the system can be $R_0 = \max\{R_0^1, R_0^2, \ldots, R_0^n\}$ [6].

3.4   Threshold Host Size to Sustain Macroparasite Infection

Microparasites are tiny pathogens such as viruses, bacteria, and some protozoa, whereas the macroparasites are typically larger pathogens such as some other kinds of protozoa and helminths. The life cycles of macroparasites typically require either vectors or other intermediate hosts to be completed. In microparasite infection, the intensity of the disease in the host (symptoms and infectiousness) is, in general, determined by whether the host is infected or not. That is, infected host has a large number of parasite loads due to high reproductive rates (multiply within their definitive host). On contrary, the number of parasites per host is an important variable in case of macroparasite infection as it may drastically vary between infected individuals. Hence, from modeling perspective, macroparasites are modeled in a different way from microparasite [22].

The basic model of macroparasite includes interaction populations of an average number of adult macroparasites within a host (P), size of host population (H), and free-living parasite larvae in the environment (W). The parameters in the model include per capita birth of hosts (b), per capita natural death rates of hosts (d), the encounter rate between host and parasite (β), and death rates of adult parasite within the host (μ; also die if their host dies by natural death or due to infection (α)). The parameters δ and α can be considered as the per capita effects of each adult parasite on host fecundity and mortality, respectively. The macroparasite aggregation in host population is modeled via a negative binomial distribution with dispersion parameter, k, which can be interpreted as decrease in fecundity in infected hosts due to higher macroparasite burden. The larval population depends on the fecundity of adult parasites (λ), larval mortality rate in the environment (γ), and the rate at which larvae are transmitted successfully to new susceptible hosts (β). Using this information, we can describe the macroparasite system [8, 11, 14] as

$$dH/dt = (b - d)H - (\alpha + \delta)P$$

$$dP/dt = \beta WH(\mu + d + \alpha)P - \alpha \frac{P^2}{H} \frac{k+1}{k}$$

$$dW/dt = \lambda P \gamma W \beta WH \tag{18}$$

In the macroparasite model, R_0 is interpreted as the number of parasites produced by an adult parasite over its life span and is given by

$$R_0 = \frac{\lambda \beta H}{(\mu + d + \alpha)(\gamma + \beta H)} \tag{19}$$

From this, we can compute the threshold number of hosts, H_T, required to sustain continuous parasite infection. Setting $R_0 = 1$, we can obtain

$$H_T = \frac{\gamma(\mu + d + \alpha)}{\beta(\lambda - (\mu + d + \alpha))} \tag{20}$$

3.5 The Reproduction Number in Social Issues

We consider a model from [16] (see Fig. 5) to describe drug and tobacco use. A similar model has also been used to study crack cocaine use (other applications are in [24]). Here, S represents the at-risk population into susceptible non-participants, D represents participants/users, and former (recovering R or arrested A) participants. Let the pressure contact rate per capita into the D user class be $\frac{\beta D}{N}$ and departure rate per capita be $\mu + \gamma$, which gives the basic reproduction number without relapse is $\frac{\beta}{\mu + \gamma}$. The relapse rate per capita given by δ adds to the reproduction number, as the average number of times that an individual visits the user class is greater than 1. Instead, a fraction $\frac{\gamma}{\mu + \gamma}$ of those who leave the user class go into the recovering class, of whom a fraction $\frac{\delta}{\mu + \delta}$ then relapses back into the user class. In all, a fraction

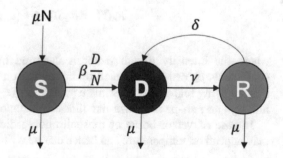

Fig. 5 The model for drug use with linear relapse. S represents the at-risk population into susceptible non-participants, D represents participants/users, and former (recovering R or arrested A) participants

$\theta = \frac{\gamma}{\mu+\gamma}\frac{\delta}{\mu+\delta}$ of users relapse. For individuals who relapse, a further fraction θ i.e, a fraction θ^2 of all users relapse a second time. Similarly, a fraction θ^3 of all users eventually relapse at least three times, etc. The average number of times a user enters the user class is therefore given by a geometric series $\sum_{n=1}^{\infty} \theta^n$, making the reproduction number of participants/users

$$R_P = \frac{\beta}{\mu+\gamma} \sum_{n=1}^{\infty} (\frac{\gamma}{\mu+\gamma}\frac{\delta}{\mu+\delta})^n \qquad (21)$$

A simple model for unhealthy eating behaviors can be also developed. Suppose the young population can be divided as: "moderately" healthy (M) and "less" healthy (L) enrolled in all schools where social programs such as standard nutrition education programs are effective at improving food choices at the per capita rate (ϕ) [26]. M-eaters can shift back to L-eaters, due to recidivism. The contagion power of L individuals would be considered successful at worsening food choices as long as the interactions between M and L lead to an increase in the number of Ls at the $\lambda = \beta L/N$ rate (where λ models M-eaters frequency-dependent interactions with L-eaters). Here, β denotes the success rate of worsening eating habits of M-eaters. For such model, the control reproduction number is given by

$$R_c(\phi) = \beta/(\phi + \mu)$$

and can be interpreted as the number of secondary conversions from M to L, over the total average time spent in the school as an L-eater (represented by $1/(\mu + \phi)$; in years).

3.6   Time-Varying Reproduction Number

The reproduction number quantity can change over time, and if it does, it is referred to as effective reproductive number. However, various social and environmental factors may influence its estimates at different times. For example, climatic variables can shape transmission via the k potential drivers, $\delta_{t,k}$, and hence, effective reproduction number can be defined as

$$R_e(t) = R_0 S(t) \prod_k (\delta_{t,k})^{\beta_k} ,$$

where the intensity of kth factor is captured by the coefficient β_k [20]. The coefficients β_k can be estimated by taking log on both sides of the equations and then applying regression to estimate parameters (e.g., absolute humidity and school holidays may shape influenza-like illness in a region).

In case of vector-borne or mosquito-borne diseases, the reproduction number may depend on temperature and hence can be written as [28]

$$R_e(T) = \frac{m(T) \cdot b \cdot c \cdot a(T)^2 \cdot e^{-\mu(T)n(T)}}{\mu(T) \cdot \gamma},$$

where $m(T)$ is the mosquito-to-human ratio as a function of temperature T, $\mu(T)$ is the mean daily mortality rate of adult mosquitoes at temperature T, b and c are human-to-mosquito and mosquito-to-human infection probabilities, $a(T)$ is the mosquito biting rate as a function of temperature, $1/\gamma$ is the average duration of infectiousness in humans, and $n(T)$ is the mean extrinsic incubation period at temperature T.

4 Application of Reproduction Number for Design of Interventions

Modeler uses different types of metrics to quantify this impact, for example, basic reproduction number (R_0), effective reproduction number (R_e), contextual control reproduction number (R_c), transmission dispersion measure (K), and critical virus replication rate (C_t). These metrics are often referred to as tipping points in their respective domains because there is a critical value of these numbers that may result in a drastic change in system behavior.

4.1 Critical Vaccination Rate

One of the critical questions in the year 2021 worldwide is to model an optimal distribution of COVID vaccine. Let us recall how smallpox was eradicated by effective vaccination, which helps bodies develop immunity to smallpox. Widespread immunization and surveillance were conducted around the world for several years, and in 1980 WHO declared smallpox has been eradicated. It is the only known human infectious disease to have achieved this distinction. If we assume that each infected person contacts N new people per infectious period, then, on average, $\frac{R_0}{N}$ of those people are likely to become infected. Now suppose V number of individuals are vaccinated among the N contacted, then $\frac{R_0}{N}(N - V)$ will be the number of new infections. In order to control the disease, this number has to be less than one. Thus, setting this number equal to 1, we will obtain the proportion $\frac{V}{N}$ of individuals need to be vaccinated in order to control the spread of the infectious disease. This gives critical vaccination rate, $V_{cri} = \frac{V}{N} = 1 - \frac{1}{R_0}$, a quantity that is needed to obtain the herd immunity threshold condition for the population. Now, if the effectiveness of vaccine is E (i.e., fraction E of vaccinated individual will be fully protected against the infection), then the critical vaccination rate, V_{cri}, will be

$$V_{cri} = \frac{V}{N} = \frac{(1 - \frac{1}{R_0})}{E} \tag{22}$$

4.2   Critical Virus Replication Rate in Diagnostic Testing

Testing is a population-level key indicator of infection spread and severity. There are different type of tests available in terms of both identification particles (e.g., live virus, antigen, and antibodies) and speed of testing (rapid test versus RT-PCR testing). In general, delay in reporting of test results is a function of a shortage of supplies for testing, shortage of protective gears for healthcare workers, patient inconvenience to take tests frequently (e.g., swabs are uncomfortable), cost associated with undertake testing for healthcare facility, differences in testing policies from one region to another, and wide spectrum of patients' clinical manifestation (e.g., asymptomatic, mild symptoms, and moderate to severe symptoms). It also has population-level implications; longer the average delay in testing, the longer it takes to control an infection from the community and higher the number of infections will be generated during the course of the outbreak. At individual levels, longer healthcare workers uninfected will take to come back to work, and patients/society will have larger productivity loss. In-house facilities usually have fast turnaround times but can handle much smaller volume. It saves time as it does not require packaging, sent back and forth between different facilities, systems processing, prioritizing, and tracking of the tests, which cuts down on red tape and improves efficiency.

The discussion on testing is relevant for not only patients but also society as a whole because weak policies in response to this may result in more suffering and societal direct and indirect costs. One such cost is a secondary number of new cases that a patient who is infected but is not tested and quarantined properly may generate. In reverse transcription polymerase chain reaction (RT-PCR) test, which is the gold standard for diagnosing COVID-19, a positive reaction is detected by accumulation of a fluorescent signal. Critical virus replication rate, C_T, is a value that indicates the number of cycles needed in the RT-PCR test to amplify viral RNA, so it can reach a detectable level. A lower C_T value, 29 or below, is a sign of high viral load. A C_T value of 30–37 indicates moderate amounts of viral load. A C_T value of 38–40 indicates minimal viral load. The high viral load may further indicate threshold for infectiousness of infected individual leading to high possibility of transmission of pathogens to susceptible, as well as trajectory of infection profile or severity of the disease in path to recovery. The C_T value may differ depending on the test kit used. Hence, the infection status of an individual is indicated by PCR test:

$$\mathbb{1}_{\mathscr{X}} = \begin{cases} +ve & \text{if } C_T \in X = [0, \, 40) \\ -ve & \text{if } C_T \in X > 40 \end{cases}$$

implying $C_T \approx 40$ is a threshold level of the PCR test to identify if the person is infected or not.

4.3   Virus Generation Rate in a Host's Immune Cell

The human immunodeficiency virus (HIV), which causes around 40,000 new HIV infections per year, has been a threat to the world population, in part because of its harmful effect on the immune system. The HIV virus typically attacks CD4+ T cells of the immune system and multiplies within them, eventually budding off and killing the host cell. Once the CD8+ T cells of the immune system get aware of the infection, they begin destroying actively infected CD4+ T cells and free virions. There is a continuous tug of war between the size of virus population and CD4+ T cells population. Higher CD4+ T cells count indicates stronger immune response to HIV, and however, the rate of production of HIV almost always overpowers the immune system. Hence, a simple mathematical model capturing the behavior of the HIV virus at a cellular level along with immune system can be readily derived [5]. The model incorporates T_K (which denotes the concentration of CD8+ T cells), T_H (which denotes the concentration of uninfected CD4+ T cells), T_I (which denotes the concentration of actively infected CD4+ T cells), and V (which denotes the concentration of free infectious virus particles). For example, the virus equation is given as follows:

$$T_K' = s_1 - \mu_K T_K + r_K T_K \left(1 - \frac{T_K}{T_{Kmax}}\right)$$

$$T_H' = s_2 - \mu_H T_H - kV T_H + r_H T_H \left(1 - \frac{T_H + T_I}{T_{Hmax}}\right)$$

$$T_I' = kV T_H - \mu_I T_I - \delta T_I T_K$$

$$V' = N\mu_H T_I - kV T_H - \mu_V V \tag{23}$$

where s_1 (and s_2) represents the rate at which the thymus supplies CD8+ T cells (and CD4+ T cells), r_K (and r_H) represents the replication rate of CD8+ T cells (and CD4+ T cells), T_{Kmax} (and T_{Hmax}) is the maximum number of CD8+ T cells (and CD4+ T cells), μ_K (and μ_H) represents the natural death rate of CD8+ T cells (and CD4+ T cells), μ_V represents the natural death rate of virus cells, δ represents the rate at which CD8+ T cells kill the actively infected CD4+ T cells, N *is the average number of free viruses produced by the death of infected CD4+ T cells*, and k represents the rate at which CD4+ T cells become infected.

Analyzing this system, we can compute critical size N_{cri} of virus generation rate (from infected CD4+ T cells bursting) required for infection to sustain

$$N_{cri} = \frac{(\mu_V + kT_H^*)(\mu_I + \delta T_K^*)}{kT_H^*} \tag{24}$$

where T_H^* and T_K^* are the components of the virus-free steady state in which V^* and T_I^* are zero.

4.4   Basic Reproduction Number in Healthcare Settings

Describing and capturing contact structure between individuals are fundamental to the spread of infectious diseases. Moreover, contact patterns in closed contacts indoor facility such as healthcare settings are highly context-specific. For example, contacts between patients and healthcare workers in a hospital are typically more frequent, longer, and of higher risk than contacts that occur in the community. Therefore, using simplifying assumptions, we can derive an expression of R_0 for the closed contacts places, interpreted average number of new cases produced in an indoor setting, as follows [31]:

$$R_0 = p \times d_C \times n_C \times d_I \qquad (25)$$

where p is the probability of transmission per minute spent in contact, d_C is the average contact duration (in minutes), n_C is the average number of contacts per person per day, and d_I is the average duration of infectivity (in days). Assuming that p and d_I are the same for individuals in the local community (L) and in healthcare settings (H), we can translate the previous expression into setting-specific R_0 as R_0^L and R_0^H. Hence, the healthcare setting-specific reproduction number may be estimated from the community-specific reproduction number and the contact pattern characteristics in both settings as follows:

$$R_0^H = R_0^L \times \frac{d_C^H \times n_C^H}{d_C^L \times n_C^L} \qquad (26)$$

5   Distribution of Reproduction Number (Capturing Variance in R_0) or K-number

These averages (i.e., reproduction numbers) are not always useful for understanding the spread of infection in the community, especially when behavior changes widely due to frequent policy shift and lockdown/opening of economy. In fact, in many infectious diseases, 80–20 rule (Pareto principle) has been observed. That is, a few infected person (20%) generated 80% of the new cases. Often, it is also observed that there is a differential incidence and/or mortality rate between regions in spite of similarities (in population size/density, age distribution, travel patterns, weather, etc.). The question that comes to our mind is how can we explain these variations in affected population? The average reproduction number may not be a good representative. The k-number or dispersion measure quantifies whether a virus spreads in a steady manner (spreading homogenously in the population) or in clusters (whereby one person infects many, all at once). This number extends the idea of reproduction number (which is an average quantity) and captures its over-dispersion behavior (super infectious on one side to noninfectious on other side). It is useful as close contact superspreading events are critical in the late stages

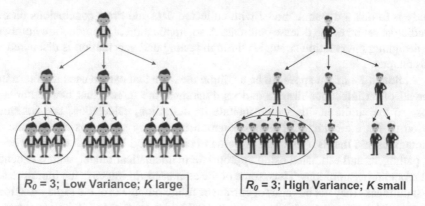

Fig. 6 The same mean estimate of R_0 but with low and high variances in R_0. An arrow indicates successful infection transmission from a person. The low and high variances in R_0 correspond to large and small K-value, respectively

of the epidemic when the virus is almost eradicated. It is derived by assuming a negative binomial offspring distribution in the branching process with a mean of R-effective (R_t or R_{eff}) and a dispersion parameter K (represents the degree of transmission heterogeneity, with lower values of K corresponding to higher variance). For example, for the 1918 influenza, $K = 1$ (40% of infected people might not pass on the virus to anybody else), whereas for the SARS, MERS, and COVID-19, $K = 0.1$ (70% infected do not infect anyone).

Consider a branching process representing a number of infected individuals in a population over time. Suppose the random variable, Z, represents the number of secondary cases caused by each infectious individual in a population. Let the offspring distribution of Z be modeled by a Negative Binomial process with mean as the population reproduction number (i.e., $Z \sim NegativeBinomial(R_0, K)$). Note that the expectation (or R_0) can take any positive real value, while Z is a non-negative integer ($0, 1, 2, 3, \ldots$). In the negative binomial distribution, smaller values of K indicate greater heterogeneity in the secondary cases [17]. Note that, if all individuals have the same infectiousness, i.e., the variance is low, then the number of secondary infections is expected to have a Poisson distribution (i.e., $K \to \infty$), and if the infectiousness is heterogeneous, the distribution is overdispersed (has a lower K) (see Fig. 6).

6   Concluding Remarks

We have introduced the concept of tipping point, basic reproduction numbers, and dispersion number and have used several examples of interest in mathematical biology. There are many different reproduction numbers that we have discussed in this chapter and are helpful in different scenarios [27]. It is used as a key predictive tool to understand and control a disease outbreak. The typical goal of a modeling study is to link a dynamic model with collected data and draw conclusions on the predictive nature of the disease outbreak. Also, application of reproduction numbers in designing intervention strategies through testing and vaccination is discussed in this chapter.

Although R_0 might appear to be a simple measure that can be used to determine spread of an infectious disease and its dynamics, the threats that new outbreaks pose to the public health, which include its definition, calculation, and interpretation of R_0, are complex. The reproduction number in general is an estimate of contagiousness that is a function of human behavior and biological characteristics of pathogens and estimated using appropriate mathematical model, which depends on modeling assumptions. Thus, some of the estimated R_0 values in the literature for past outbreaks may not be valid for current or future outbreaks of the same infectious disease. Therefore, reproduction number or tipping point needs to be constructed and interpreted based on the assumptions and context in which it is used.

The spread of information, disease, social problem, policy responses, etc. is open to different interpretation due to access to varied levels of evidence [24]. Individual's or population's decisions can be tied to the narratives that develop over the propagation of an issue faced by the public. Narratives are the stories of emergence, spread, and control of an issue of a population in its social, cultural, and economic contexts [35]. Narratives have played a key role in the dissemination of an issue and effectiveness of responses to it. The dynamics of an issue are often governed by policy narratives as well as influenced by policy development and implementation. Sometimes organization self-interests or simplification of a policy for the understanding of the general public (overlooking essential facts) can lead to success of the policy in claiming the narrative, so that it can be able to rally the people to its side, irrespective of long-term consequences. On the other hand, shifting narrative can lead to arguments and even widespread violence and thus illuminating population with new and better public perspectives. There is a trade-off in the level of variations of narratives to an issue. Narratives come with its own risk level and are governed by its respective (often hidden) reproduction number. The conceptualization and interpretation of this threshold quantity, reproduction number, of an issue is an important factor but only when it can be properly evaluated during its natural course, and all the relevant datasets are analyzed. Rushed out or ignored efforts to compute reproduction numbers can result in misleading or uninformed policy decisions.

Research Project 1 Consider the usual SEIR model

$$s' = \Pi - vS - \gamma SI \tag{27}$$

$$E' = \gamma SI - (v + k)E \tag{28}$$

$$I' = kE - (v + \rho)I \tag{29}$$

(continued)

$$R' = \rho I - \nu R \tag{30}$$

where individuals progress from compartment E to I at a rate k and develop immunity at a rate ρ, natural mortality claims individuals at a rate μ, and there is a constant recruitment, Π, of susceptible individuals. The basic reproduction number R_0 is calculated as $R_0 = \frac{k\gamma P_0}{(\nu+k)(\nu+\rho)}$, where $P_0 = \frac{\Pi}{\nu}$. Interpret the basic reproduction number formula, and verify that the disease-free equilibrium is (locally asymptotically) stable for $R_0 < 1$ and unstable for $R_0 > 1$ (note that this can be done by computing Jacobian of the system and its eigen values). Also, now if the disease-induced death is included in the above model, then how the formula of the basic reproduction number changes? [1]

Research Project 2 The Ebola, viral infectious disease, has erupted occasionally in Africa continent. Ebola is a highly contagious and deadly disease with seriously infected individual die in about 10 days after infection. Interventions like quarantine (as well as isolation) of patients is one of the best measures to control the spread of the Ebola. Construct a model for the spread of Ebola that includes quarantine compartment, which consists of individuals who are temporary isolated from the rest of the population. Find the expression of the basic reproduction number [1].

Research Project 3 Consider a model where the recovered class (V) is now the immune class with immunity due to both recovery and vaccination of susceptible individuals:

$$S'(t) = \mu N - \beta \frac{SI}{N} - (\mu + \phi)S$$

$$I'(t) = \beta \frac{SI}{N} - (\mu + \gamma)I$$

$$V'(t) = \gamma I + \phi S - \mu V$$

1. Show that $\frac{dN}{dt} = 0$, where $N = S + I + V$. What does this result imply?
2. Discuss why it is enough to study the first two equations.

(continued)

City in WI/Dates	April 4	April 5	April 6	April 7	April 8	April 9	April 10
Franklin	15	15	18	19	21	24	27
S. Milwaukee	14	16	17	20	21	23	28

https://county.milwaukee.gov/EN/COVID-19

Fig. 7 The data in the table represents the number of reported COVID-19 infected people in two different cities in the USA from April 4 to April 10. Data source https://county.milwaukee.gov/EN/COVID-19

3. Let $R_0(\phi)$ be R_0 when $\phi \neq 0$ and $R_0(0)$ be R_0 when $\phi = 0$. Compute $R_0(\phi)$. What is the value of $R_0(0)$? Compare $R_0(\phi)$ with $R_0(0)$.
4. Using the formula of the $R_0(\phi)$, compute the critical vaccination rate (see Sect. 4.1 for details).

Research Project 4 Consider the data in Fig. 7 [3] and the note at the end of Project 4. Perform following steps:

1. Use Excel or Google Sheets to graph a scatter plot with y-axis (the number of infected) and x-axis (dates).
2. Write down your observations about the trends in the graph. Does the early trend follow an exponential growth curve? Why or Why not?
3. Assume the data for the two cities follow exponential function, say, $I(t) = I(0)e^{\lambda t}$. Estimate λ for the two cities. (Hint: take log of equation on both sides, and you will get equation of line. Also, take log of data in the table. Then, fit line to log of data in Excel.)
4. Estimate a basic reproduction number R_0 for both the cities, considering the SIR model (Hint: the SIR model have two formulas for R_0: (i) $R_0 = \beta/\alpha$ and (ii) $R_0 = 1 + \lambda/\alpha$, where β is transmission rate, α is recovery rate, and λ is the rate of growth of exponential rise in cases. Estimate β for the two cities using the two formulas of R_0 and step 3 above, assuming that $\alpha = 1/14 = 0.0714$, which is 14 days infectious period.).
5. Find COVID-19 data (with 10 time points) similar to the one shown in Fig. 7 corresponding to another USA city or a country, and then use above four steps to estimate R_0 from that new chosen city/country.

(continued)

(**Note [18]:** *The initial exponential growth rate of an outbreak is, by itself, an important measure for the speed of spread of an infectious disease. The rate equal to zero is a disease threshold which is equivalent to the basic reproduction number $R_0 = 1$. The disease can invade a population if the growth rate is positive and cannot invade (with a few initially infectious individuals) if it is negative. Consider the SIR epidemic model, and its disease-free equilibrium is $(1, 0, 0)$, representing a completely susceptible population. Using linearization about disease-free equilibrium (that is, using largest eigen value of the linearization), we can approximate the growth rate. Then, using standard $R_0 = \beta/\alpha$ formula for the model, we can estimate the exponential growth rate as $\lambda = \beta - \alpha$. If, for example, α can be estimated independently of λ, then the basic reproduction number is $R_0 = 1 + \lambda/\alpha$.*)

References

1. Linda JS Allen, Fred Brauer, Pauline Van den Driessche, and Jianhong Wu. *Mathematical epidemiology*, volume 1945. Springer, 2008.
2. Barry C Arnold. Pareto distribution. *Wiley StatsRef: Statistics Reference Online*, pages 1–10, 2014.
3. Cabral Balreira, Casey Hawthorne, Grace Stadnyk, Zeynep Teymuroglu, Marcella Torres, and Joanna Wares. Resources for supporting mathematics and data science instructors during covid-19. *Letters in Biomathematics*, 8(1):49–83, 2021.
4. Samuel A Bozzette, Rob Boer, Vibha Bhatnagar, Jennifer L Brower, Emmett B Keeler, Sally C Morton, and Michael A Stoto. A model for a smallpox-vaccination policy. *New England Journal of Medicine*, 348(5):416–425, 2003.
5. Antonio Buenrostro, Katie Diaz, CP Gonzáles, and Magdaliz Gorritz. HIV and its impact on the infant immune system. *MTBI technical report*, 2004.
6. Britnee Crawford and Christopher M Kribs-Zaleta. The impact of vaccination and coinfection on HPV and cervical cancer. *Discrete & Continuous Dynamical Systems-B*, 12(2):279, 2009.
7. Hans De Sterck, John Lang, and ORCAS Seminar. The Arab Spring: A simple compartmental model for the dynamics of a revolution. *Dynamics*, (1/41), 2013.
8. Andrew P Dobson and Peter J Hudson. Regulation and stability of a free-living host-parasite system: Trichostrongylus tenuis in Red Grouse. II. Population models. *Journal of Animal Ecology*, pages 487–498, 1992.
9. Malcolm Gladwell. *The tipping point: How little things can make a big difference*. Little, Brown, 2006.
10. Simon Gubbins, Simon Carpenter, Matthew Baylis, James LN Wood, and Philip S Mellor. Assessing the risk of bluetongue to UK livestock: uncertainty and sensitivity analyses of a temperature-dependent model for the basic reproduction number. *Journal of the Royal Society Interface*, 5(20):363–371, 2008.
11. Barbara A Han, Andrew W Park, Anna E Jolles, and Sonia Altizer. Infectious disease transmission and behavioural allometry in wild mammals. *Journal of Animal Ecology*, 84(3):637–646, 2015.
12. Jane M Heffernan, Robert J Smith, and Lindi M Wahl. Perspectives on the basic reproductive ratio. *Journal of the Royal Society Interface*, 2(4):281–293, 2005.
13. J Holland. Notes on r0 department of anthropological sciences. *Stanford University, Department of Anthropological Sciences, Building*, 360:94305–2117, 2007.

14. Peter J Hudson, David Newborn, and Andrew P Dobson. Regulation and stability of a free-living host-parasite system: Trichostrongylus tenuis in Red Grouse. I. Monitoring and parasite reduction experiments. *Journal of animal ecology*, pages 477–486, 1992.

15. William Ogilvy Kermack and Anderson G McKendrick. A contribution to the mathematical theory of epidemics. *Proceedings of the royal society of London. Series A, Containing papers of a mathematical and physical character*, 115(772):700–721, 1927.

16. Christopher M Kribs-Zaleta. Sociological phenomena as multiple nonlinearities: MTBI's new metaphor for complex human interactions. *Mathematical Biosciences & Engineering*, 10(5&6):1587, 2013.

17. James O Lloyd-Smith, Sebastian J Schreiber, P Ekkehard Kopp, and Wayne M Getz. Super-spreading and the effect of individual variation on disease emergence. *Nature*, 438(7066):355–359, 2005.

18. Junling Ma. Estimating epidemic exponential growth rate and basic reproduction number. *Infectious Disease Modelling*, 5:129–141, 2020.

19. Angus McLure and Kathryn Glass. Some simple rules for estimating reproduction numbers in the presence of reservoir exposure or imported cases. *Theoretical population biology*, 134:182–194, 2020.

20. C Jessica E Metcalf, Katharine S Walter, Amy Wesolowski, Caroline O Buckee, Elena Shevliakova, Andrew J Tatem, William R Boos, Daniel M Weinberger, and Virginia E Pitzer. Identifying climate drivers of infectious disease dynamics: recent advances and challenges ahead. *Proceedings of the Royal Society B: Biological Sciences*, 284(1860):20170901, 2017.

21. Jeffrey D Morenoff and Robert J Sampson. Violent crime and the spatial dynamics of neighborhood transition: Chicago, 1970–1990. *Social forces*, 76(1):31–64, 1997.

22. Anuj Mubayi. Inferring patterns, dynamics, and model-based metrics of epidemiological risks of neglected tropical diseases. In *Handbook of statistics*, volume 37, pages 155–183. Elsevier, 2017.

23. Anuj Mubayi, Marlio Paredes, and Juan Ospina. A comparative assessment of epidemiologically different cutaneous leishmaniasis outbreaks in Madrid, Spain and Tolima, Colombia: an estimation of the reproduction number via a mathematical model. *Tropical medicine and infectious disease*, 3(2):43, 2018.

24. Anuj Mubayi, Jeff Sullivan, Jason Shafrin, Oliver Diaz, Aditi Ghosh, Anamika Mubayi, Olcay Akman, and Phani Veeranki. Battling epidemics & disparity with modeling. *Letters in Biomathematics*, 7(1):105–110, 2020.

25. Antoine Perasso. An introduction to the basic reproduction number in mathematical epidemiology. *ESAIM: Proceedings and Surveys*, 62:123–138, 2018.

26. Muntaser Safan, Anarina L Murillo, Devina Wadhera, and Carlos Castillo-Chavez. Modeling the diet dynamics of children: The roles of socialization and the school environment. *Letters in biomathematics*, 5(1):275, 2018.

27. Ram Singh, Naveen Sharma, and Aditi Ghosh. Modeling assumptions, mathematical analysis and mitigation through intervention. *Letters in Biomathematics*, 6(2):1–19, 2019.

28. Amir S Siraj, Rachel J Oidtman, John H Huber, Moritz UG Kraemer, Oliver J Brady, Michael A Johansson, and T Alex Perkins. Temperature modulates dengue virus epidemic growth rates through its effects on reproduction numbers and generation intervals. *PLoS neglected tropical diseases*, 11(7):e0005797, 2017.

29. David L Smith and F Ellis McKenzie. Statics and dynamics of malaria infection in anopheles mosquitoes. *Malaria journal*, 3(1):1–14, 2004.

30. Nils Chr Stenseth, Herwig Leirs, Anders Skonhoft, Stephen A Davis, Roger P Pech, Harry P Andreassen, Grant R Singleton, Mauricio Lima, Robert S Machang'u, Rhodes H Makundi, et al. Mice, rats, and people: the bio-economics of agricultural rodent pests. *Frontiers in Ecology and the Environment*, 1(7):367–375, 2003.

31. Laura Temime, Marie-Paule Gustin, Audrey Duval, Niccolò Buetti, Pascal Crepey, Didier Guillemot, Rodolphe Thiébaut, Philippe Vanhems, Jean-Ralph Zahar, David RM Smith, et al. A conceptual discussion about the basic reproduction number of severe acute respiratory syndrome coronavirus 2 in healthcare settings. *Clinical Infectious Diseases*, 72(1):141–143, 2021.

32. Joanne Turner, Roger G Bowers, and Matthew Baylis. Two-host, two-vector basic reproduction ratio (r 0) for bluetongue. *PloS one*, 8(1):e53128, 2013.
33. Pauline Van den Driessche. Reproduction numbers of infectious disease models. *Infectious Disease Modelling*, 2(3):288–303, 2017.
34. Pauline Van den Driessche and James Watmough. Reproduction numbers and sub-threshold endemic equilibria for compartmental models of disease transmission. *Mathematical biosciences*, 180(1-2):29–48, 2002.
35. Priscilla Wald. *Contagious: cultures, carriers, and the outbreak narrative*. Duke University Press, 2008.
36. Shi Zhao, Salihu S Musa, Jay T Hebert, Peihua Cao, Jinjun Ran, Jiayi Meng, Daihai He, and Jing Qin. Modelling the effective reproduction number of vector-borne diseases: the yellow fever outbreak in Luanda, Angola 2015–2016 as an example. *PeerJ*, 8:e8601, 2020.

Application of Mathematics to Risk and Insurance

Kumer P. Das

Abstract

A merit rating system known as bonus-malus system (BMS) is popular in many European, Asian, and South American countries. Compared to a flat-rate system, a bonus-malus system is better able to reflect risk levels in insurance premium as it depends solely on the current class of the driver and the number of claims occurred during the period. Under certain conditions, the successive bonus classes for a policy form a Markov chain with a transition matrix. As insurance companies are primarily interested in the stabilization of the mean and variance of premium level, it is important to understand the steady-state or stationary condition of a BMS. This study sets out to accomplish the following: providing a mathematical foundation of BMS, describing the big three claim distributions, providing a brief descriptions of number of several BMSs, and calculating stationary distributions.

Suggested Prerequisites *Linear Algebra, Introduction to Inferential Statistics, Probability Theory.*

K. P. Das (✉)
University of Louisiana at Lafayette, Lafayette, LA, USA
e-mail: kumer.das@louisiana.edu

1   Introduction

Unlike in life insurance where the use of a priori variables such as age, sex, occupations, health characteristics, etc. is very common, in auto insurance the best predictor of the future number of claim is a driver's past claim behavior. Bonus-malus (Latin for good-bad) systems are very common in auto insurance in Europe and some parts of Asia and South America. This is a merit rating system containing several classes and adjusts the premium of a customer according to the customer's claim history. Compared to a flat-rate system, a BMS reflects risk levels in insurance premium better [10]. In each period (usually a year) all policyholders are divided into a finite number of classes C_i $(i = 1, 2, \ldots, n)$ and their insurance premium depends only on the class they belong to. Claims are penalized by malus points where a driver goes up a certain number of classes each time a claim is filed by the driver. On the other hand, each claim-free year is rewarded by bonus points where the driver goes down usually one class. BMS forms a Markov chain process under certain conditions because a BMS requires the system to be Markovian, or "memory-less," which states that the knowledge of the present class and the number of accidents during the policy term (for example, a year) are sufficient to determine the class of the next year [5].

The rest of the paper are organized as follows: definitions, notations, and background information on BMS are provided in Sect. 2. A brief description of three commonly used claim frequency models is provided in Sect. 3. Section 4 provides a few examples of BMSs used in several countries around the world. Section 5 discusses stationary distribution of BMS with a numerical example.

2   Definitions and Notation

Consider a BMS with n classes. As [7,9] and [2] stated, an insurance company uses a BMS if the following assumptions are valid:

- There are a finite number of classes, denoted by C_i or simply i $(i = 12, \ldots, n)$, in a policy so that the premium for a given period (usually, a year) depends solely on the class.
- The class for a given period solely depends on the class for the previous period and the number of claims occurred during that period.

A BMS is determined by the following elements:

- The initial class, usually denoted by C_{i_o},
- The premium level $(b_1, b_2, \ldots, b_n)$,
- The number of claims is k with $k \in Z^*$,
- The rules that determine the transition from one class to another when the number of claims in a given period is known. These rules are known as transition rules

and are denoted as $T_k(i) = j$ if the policy is transferred from class C_i into class C_j when k claims are reported.

The transformation T_k can be written in the form of a matrix

$$T_k = (t_{ij}^{(k)}), \tag{1}$$

where $t_{ij}^{(k)} = 1$ if $T_k(i) = j$ and 0 otherwise. Thus, T_k is a binary matrix which is also known as a Boolean matrix. The probability of a policy moving from C_i into C_j in one period is denoted by $P_{ij}(\lambda)$ where λ is the expected number of claims per period for the policy. Mathematically,

$$P_{ij}(\lambda) = \sum_{k=0}^{\infty} p_k(\lambda) t_{ij}^{(k)},$$

where $p_k(\lambda)$ is the probability that a policyholder with claim frequency λ has k claims in a policy period. Since $P_{ij}(\lambda)$ is a probability, we have,

- $P_{ij}(\lambda) \geq 0$,
- $\sum_{j=1}^{n} P_{ij}(\lambda) = 1$.

With above notations, we define the matrix

$$M(\lambda) = (P_{ij}(\lambda)) = \sum_{k=0}^{\infty} p_k(\lambda) T_k$$

as the transition matrix of a Markov chain. A first order Markov chain is a stochastic process in which the future development depends only on the present state but not on the history of the processes. In short, a sequence of random variables $X_1, X_2, \ldots$ form a Markov chain if

$$Pr(X_{n+1} = j | X_1 = i_1, X_2 = i_2, \ldots, X_n = i) = Pr(X_{n+1} = j | X_n = i).$$

A Markov chain is a process without memory because given the present state, the future and past states are independent. States of the Markov chain are the different classes of a BMS. Since the knowledge of the present class and the number of claims for the time period are enough to determine the next period's class we can see why a BMS may turn into a Markov chain. It should be noted that a BMS may be non-Markovian too as it forces insurance companies to memorize the recent claim history. Another important characteristics of a BMS is known as the *spread factor* defined by [11] as the ratio between the highest premium b_n where n is the number of class and the lowest premium b_1.

Consider a Markov chain example with a rat in a maze with six cells (indexed 1–6).The example is motivated by a lecture note [1].The outside of the maze is *freedom* (indexed by 0). The *freedom* can be reached only via cell 6.The rat starts initially in any given cell and in every step move to another cell. The ultimate goal for the rat is to reach to *freedom*.It is assumed that moving to an accessible cell is equally likely and independent of the past. Let X_n denotes the cell visited right after the nth move, then $\{X_n\}$ constitutes a Markov chain with state space $S = \{0, 1, 2, 3, 4, 5, 6\}$. The transition matrix of this example is given by

$$
\mathbf{P}=
\begin{array}{c}
 \\
0 \\
1 \\
2 \\
3 \\
4 \\
5 \\
6
\end{array}
\begin{pmatrix}
\phantom{\frac{1}{2}}0 \ \ 1 \ \ 2 \ \ 3 \ \ 4 \ \ 5 \ \ 6\phantom{\frac{1}{2}} \\
1 \ \ 0 \ \ 0 \ \ 0 \ \ 0 \ \ 0 \ \ 0 \\
0 \ \ 0 \ \ \frac{1}{2} \ \ \frac{1}{2} \ \ 0 \ \ 0 \ \ 0 \\
0 \ \ \frac{1}{2} \ \ 0 \ \ 0 \ \ \frac{1}{2} \ \ 0 \ \ 0 \\
0 \ \ \frac{1}{3} \ \ 0 \ \ 0 \ \ \frac{1}{3} \ \ \frac{1}{3} \ \ 0 \\
0 \ \ 0 \ \ \frac{1}{3} \ \ \frac{1}{3} \ \ 0 \ \ 0 \ \ \frac{1}{3} \\
0 \ \ 0 \ \ 0 \ \ \frac{1}{2} \ \ 0 \ \ 0 \ \ \frac{1}{2} \\
\frac{1}{3} \ \ 0 \ \ 0 \ \ 0 \ \ \frac{1}{3} \ \ \frac{1}{3} \ \ 0
\end{pmatrix}
$$

The matrix $\mathbf{P} = (P_{ij})$ for $i, j \in S$ is known as a one-step transition matrix with the following properties:

- It is a square matrix;
- For each $i \in S$, $\sum_{j \in S} P_{ij} = 1$.

As the rat will not go back to the maze after it reaches the state of freedom, the state 0 is called an absorbing state. Very often, we want to determine the expected number of moves, $E(\tau_{i,0})$, required for the rat until it reaches the state 0 given that the rat starts initially in cell i. Mathematically, the number of moves required for the rat to reach freedom (state 0) when starting in cell i is,

$$\tau_{i,0} = min\{n \geq 0 : X_n = 0 | X_0 = i\}.$$

Our objective is to compute $E(\tau_{i,o})$ by conditioning on X_1. Say, we are interested to calculate $E(\tau_{4,o})$ which is the expected number of moves the rat needs until it reaches freedom given that it starts initially in cell 4. Whenever the rat is in cell 4, it moves next (regardless of its past) into cell 2 or 3 or 6 with equal probability ($\frac{1}{3}$).By notation,

$$P(X_1 = 2|X_0 = 4) = P(X_1 = 3|X_0 = 4) = P(X_1 = 6|X_0 = 4) = 1/3.$$

By Markovian property, we have,

$$E[\tau_{4,0}] = E[\tau_{4,0}|X_1 = 2]P(X_1 = 2|X_0 = 4) + E[\tau_{4,0}|X_1 = 3]P(X_1 = 3|X_0 = 4)$$
$$+ E[\tau_{4,0}|X_1 = 6]P(X_1 = 6|X_0 = 4)$$

$$= (1 + E[\tau_{2,0}])(\frac{1}{3}) + (1 + E[\tau_{3,0}])(\frac{1}{3}) + (1 + E[\tau_{6,0}])(\frac{1}{3})$$

$$= 1 + \frac{1}{3}E[\tau_{2,0}] + \frac{1}{3}E[\tau_{3,0}] + \frac{1}{3}E[\tau_{6,0}].$$

Exercise 1 Write similar equations for $E[\tau_{1,0}]$, $E[\tau_{2,0}]$, $E[\tau_{3,0}]$, $E[\tau_{5,0}]$ and $E[\tau_{6,0}]$ as Equation $E[\tau_{4,0}]$ above. We should have a system of six linear equations with six unknowns. Solve the system of equations to get $E[\tau_{i,0}]$ for $i = 1, \ldots, 6$.

3 The Big Three Claim Frequency Models

Counting distributions are used to model claim frequency. The three commonly used such distributions are: Poisson, binomial, and negative binomial. In practice, these claim frequency models are combined with the claim severity models to provide a compound risk model. We provide short descriptions of these three distributions in the following sections.

3.1 Poisson Distribution

A random variable X has a Poisson distribution if its probability mass function is:

$$P(X = k) = p_k = \frac{e^{-\lambda}\lambda^k}{k!}, \tag{2}$$

for some positive constant λ and for $k = 0, 1, 2, \ldots$ where λ is the only parameter of the Poisson distribution which happens to be the mean and variance of the distribution. The Poisson distribution is used to model the number of events occurring within a given time interval. For example, this distribution can be used to make forecasts about the number of customers entering in a store on a certain time interval. This is also very useful for smart order routers and algorithmic trading. One of the primary drawbacks of using a Poisson based distribution is over-dispersion which describes the observation that variation is higher than would be expected. Poisson distribution does not have a parameter to fit variability of the observation. In fact, the variance increases with the mean. For a observed data set, the expected value can be $E(X) = 10$ whereas the variance can be much higher (>10), in this case we say the data are over-dispersed and Poisson distribution cannot be suggested.

Exercise 2 The probability generating function (pgf) of a Poisson distributed variable X is

$$P_X(z) = e^{\lambda(z-1)} \tag{3}$$

for all $z \in \mathbb{R}$. a discrete distribution such as Poisson, the pgf is a power series representation (the generating function) of the probability mass function (as defined in Eq. 2) of the random variable. Using the pgf in Eq. 3, show the followings:

- The mean of X is λ,
- The variance of X is λ,
- The skewness of X is $\frac{1}{\sqrt{\lambda}}$,
- The kurtosis of X is $3 + \frac{1}{\lambda}$.

3.2 Binomial Distribution

The binomial distribution is a discrete probability distribution of the number of success (k) in a sequence of independent experiments (n) with probability of success (p). The probability mass function of a binomial random variable (X) is defined as

$$P(X = k) = p_k = \binom{n}{k} p^k q^{n-k},$$

where n is the number of trials, p is the probability of success on a single trial and $q = 1 - p$. There are two parameters (n and p) in a binomial distribution. The mean of a binomial distribution is defined as np and the variance is npq. As both p and q are probabilities, the binomial distribution requires that the variance is smaller than the mean (i.e., $np > npq$). Thus, the distribution is suitable candidate for modeling frequency for the situations where sample mean is larger than the sample variance. The probability generating function for a binomial random variable X is

$$P_X(z) = (1 - p + pz)^n.$$

Exercise 3 Show that the binomial distribution $X \sim \text{Bin}(n, p)$, can be approximated by the Poisson $X \sim \text{Poi}(np)$ when np is small.

3.3 Negative Binomial Distribution

As the name suggests, the negative binomial arises from the same probability experiment that generates the binomial distribution. However, in negative binomial setting we do not observe the outcomes in a fixed number of trials (n) as we do in binomial distribution rather we observe the trials until r number of success have occurred. The distribution was introduced by Montmort in 1714 as the distribution of the number of trials needed in an experiment to obtain a given number of successes. A random variable X has a negative binomial distribution with parameters p and r if its probability mass function is

$$P(X = k) = p_k = \binom{k + r - 1}{r - 1} p^r q^k, \tag{4}$$

with r being a positive integer and $k \in \{0, 1, 2, 3, \ldots\}$. The distribution provides the probability of $r - 1$ successes and k failures in $k + r - 1$ trials, and a success on the $(k + r)th$ trial. This distribution provides an excellent alternative to other discrete distributions where the sample variance is greater than the sample mean. For example, in the presence of Poisson over-dispersion for count data, negative binomial can provide a better model. It must be noted that the negative binomial converges to the Poisson when $r \to \infty$ and $p \to 0$, while the mean is kept constant. A more detailed discussion on the use of negative binomial distribution in modeling the distribution of aggregate claims can be found at [3]. We simplified the recursion formula proposed by Lemaire [6] to compute negative binomial probabilities

$$p_{k+1} = \frac{(k + r)q}{k + 1} p_k,$$

with

$$p_0 = p^r$$

Exercise 4 For a negative binomial random variable X as defined in Eq. 4, find the mean and variance of X.

Research Project 1 Estimation of the parameters of a probability distribution is important for various reasons. There are various methods, both graphical and numerical, for estimating the parameters. Use the method of moments and maximum likelihood to estimate the parameters of Poisson and negative binomial distribution defined in Eqs. 2 and 4, respectively. For detail information on estimation methods see [8].

4   Examples of BMS

Bonus-malus system varies from country to country primarily because of the possible variations in the number of classes in the system, transition rules and the starting class. Even within Europe, there are a number of variations as described by [4] where an analysis of BMSs for countries in Central and Eastern Europe has been presented. Moreover, based on a number of factors (for example, legislative action, conversion system) BMS in majority of the countries change frequently.

4.1   The Switzerland BMS

Switzerland, Belgium, France, Luxemburg, and Germany have very extensive BMSs. BMS in these countries have more than 20 classes and have specific transition rules [13]. The Swiss BMS has 22 classes and known as one of the toughest system in the world. The decision to enforce a strong BMS was probably influenced by the fact that Swiss insurers are only allowed to use one a priori classification variable as well as a deductible for young drivers [6]. The transition rules were as follows:

- In each claim-free year a one-class bonus is given,
- Each claim is penalized by three classes.

Moreover, there was a special bonus rule by which a policyholder with four consecutive claim-free years might not be above class 14.

4.2   The Belgian BMS

The Belgian BMS has been changed over the years. As from 2004, Belgian companies have started to exercise complete freedom of using their own BMS. In this section we will describe the old Belgian BMS which had 23 classes. The transition rules were as follows:

- Business users enter the system in class 14 whereas commuters and pleasure users enter in class 11,
- In each claim-free year a one-class bonus is given,
- Each claim is penalized by five classes.

Moreover, there was a special bonus rule by which a policyholder with four consecutive claim-free years might not be above class 14. With this rule, after a claim-free year a Belgian customer in class 18 will move to 14 or 17, depending on the number of consecutive claim-free years earned before. This restriction makes the Belgian BMS non-Markovian.

4.3 The Brazilian BMS

Brazil has one of the simplest BMS in the world. The policyholders are subdivided into just seven classes (i.e., $n = 7$). The punishment rate is one class per claim, the reward rate is also one class per claim-free period. New policy holders start in class 7. The highest premium (b_n) is 100 and the lowest premium (b_1) is 65. Thus, the *spread factor* is 1.54 or (100/65). The transition rules are presented in Table 1 as described in [5] which states that after 1 claim a policy moves to the higher class. From Eq. 1 the transformation matrix for Brazilian BMS can be developed. For example, after 1 claim we write $T_1(1) = 2, T_1(2) = 3, T_1(3) = 4, T_1(4) = 5, T_1(5) = 6$, and $T_1(\geq 6) = 7$. Also, $t_{12}^1 = 1$ since $T_1(1) = 2$ and $t_{14}^1 = 0$ since $T_1(1) = 2$. In other words, if a driver is in class 1, there are only two possible classes that the driver can be in for the next time period: class 2 if the driver makes 1 or more claim(s) or class 1 if the driver makes 0 claims.

Duplan et al. [5] developed a transition matrix for the Brazilian BMS as displayed here in Table 2 where p_0 is the probability having 0 claim in a claim period, p_1 is the probability of 1 claim in a claim period, and so on.

Table 1 Brazilian bonus-malus system

Class	Premium	Claims	Class After ---> 0	1	2	3	4	5	≥ 6
7	100		6	7	7	7	7	7	7
6	90		5	7	7	7	7	7	7
5	85		4	6	7	7	7	7	7
4	80		3	5	6	7	7	7	7
3	75		2	4	5	6	7	7	7
2	70		1	3	4	5	6	7	7
1	65		1	2	3	4	5	6	7

Table 2 Brazilian BMS transition matrix

7		6	5	4	3	2	1
7	$(1 - p_0)$	p_0	0	0	0	0	0
6	$(1 - p_0)$	0	p_0	0	0	0	0
5	$(1 - p_0 - p_1)$	p_1	0	p_0	0	0	0
4	$(1 - p_0 - p_1 - p_2)$	p_2	p_1	0	p_0	0	0
3	$(1 - p_0 - p_1 - p_2 - p_3)$	p_3	p_2	p_1	0	p_0	0
2	$(1 - p_0 - p_1 - p_2 - p_3 - p_4)$	p_4	p_3	p_2	p_1	0	p_0

Table 3 The Hong Kongese bonus-malus system

Class	Premium	Class After —> Claims	0	1	≥ 2
6	100		5	6	6
5	80		4	6	6
4	70		3	6	6
3	60		2	6	6
2	50		1	4	6
1	40		1	3	6

4.4   The Japanese BMS

The Japanese BMS has already been revised multiple times. The current BMS was introduced in 2012 which has subclasses that are no-claim and claimed subclasses for each BMS class [12].This is an unusual BMS and thus we describe the former Japanese BMS (2009 BMS) in this study. There are 20 classes in the 2009 Japanese BMS. A policyholder starts at class 6 (or class 7). The punishment rate is 3 classes per claim and the reward rate is 1 class per claim-free period of time. For example, a policyholder in class 10 who has claimed one claim goes down to class 7 and the policy holder goes up to class 11 if the policyholder does not claim.

4.5   The Hong Kong BMS

The Hong Kong BMS is one of the simplest systems in the world with only 6 classes. Its punishment rate is 2 or 3 classes on first claim and all discounts are lost for more than 1 claim. The reward rate is 1 class per claim-free period of time. The starting level for this BMS is premium 100 or class 6 (Table 3).

Table 4 Hong Kongese
BMS transition matrix

	6	5	4	3	2	1
6	$1 - p_0$	p_0	0	0	0	0
5	$1 - p_0$	0	p_0	0	0	0
4	$1 - p_0$	0	0	p_0	0	0
3	0	$1 - p_0$	0	0	p_0	0
2	0	0	$1 - p_0$	0	0	p_0
1	0	0	0	$1 - p_0$	0	p_0

5 Stationary Distribution of a BMS

Lemaire [6] provided a structure of claim frequency model where the portfolio is considered to be heterogeneous and all policyholders have constant but unequal underlying risks to have an accident. Moreover, the number of claims can be distributed as one of the three claim distributions discussed in Section 3. We will use Poisson distribution for the description of the number of claims to calculate the stationary distribution of a BMS; however, the methodology works for any other distribution. In theory, most BMS reach their steady-state or stationary condition after an infinite number of years, when the class probabilities have converged to the stationary distribution [6]. However, in practice, insurance companies are usually focused on the stabilization of the mean and variance premium level and that is why stationarity can loosely be interpreted as the mean and variance premium level being constant [6]. Mathematically, to solve for the stationary distribution, we solve the equation:

$$\mu X = \mu, \tag{5}$$

where X is the transition matrix as described in Tables 2 and 4. Here, μ is the stationary distribution and is a $1 \times n$ matrix with n being the number of classes. For Hong Kongese BMS, the setup of the stationary distribution can be written as follows:

$$[a_6, a_5, a_4, a_3, a_2, a_1] * X = [a_6, a_5, a_4, a_3, a_2, a_1]. \tag{6}$$

5.1 Numerical Examples

In the example below, we will consider Poisson distribution for the number of claims. We will also consider 5% as our claim frequency. The rationale for this choice depends on the fact that claim frequencies in most countries tend to be well under 10% based on a report published by Milliman in 2017 [4].

As the elements of a stationary distribution's sum equal to 1, we include this fact with Eq. 6 to form a system of equations. For the Brazilian BMS the system of equations can be written as follows as described in [5]:

Table 5 Different probabilities based on a 5% claim frequency

k(# of claims)	p_k	Poisson distributed values
0	p_0	0.95123
1	p_1	0.04756
2	p_2	0.00119
3	p_3	0.0

Table 6 Stationary probabilities based on a 5% claim frequency

Stationary probabilities (a_k)	Poisson distributed values
a_1	0.9474
a_2	0.0486
a_3	0.0037
a_4	0.0003
a_5	0
a_6	0
a_7	0

$$a_7 = a_7(1 - p_0) + a_6(1 - p_0) + a_5(1 - p_0 - p_1) + a_4(1 - p_0 - p_1 - p_2)$$
$$+ a_3(1 - p_0 - p_1 - p_2 - p_3) + a_2(1 - p_0 - p_1 - p_2 - p_3 - p_4)$$
$$+ a_1(1 - p_0 - p_1 - p_2 - p_3 - p_4 - p_5)$$
$$a_6 = a_7 p_0 + a_5 p_1 + a_4 p_2 + a_3 p_3 + a_2 p_4 + a_1 p_5$$
$$a_5 = a_6 p_0 + a_4 p_1 + a_3 p_2 + a_2 p_3 + a_1 p_4$$
$$a_4 = a_5 p_0 + a_3 p_1 + a_2 p_2 + a_1 p_3$$
$$a_3 = a_4 p_0 + a_2 p_1 + a_1 p_2$$
$$a_2 = a_3 p_0 + a_1 p_1$$
$$a_1 = a_2 p_0 + a_1 p_0$$
$$1 = a_1 + a_2 + a_3 + a_4 + a_5 + a_6 + a_7.$$

We can solve the above system of equations which has seven unknowns and eight equations. The objective is to obtain the probabilities for the stationary distribution. Any mathematical software such as Maple, MATLAB, Python, or R can also be used to solve the systems of equations. In Appendix the solutions of all stationary distributions in terms of probabilities have been reported. For the Brazilian BMS with claim frequency of 5% (Table 5), the Poisson claim model provides the following stationary probabilities: By definition, the stationary distribution is the left eigenvector of the transition matrix corresponding to the eigenvalue 1 which remains unchanged in the Markov chain as time progresses. For example, $a_1 = 0.9474$ indicates that 94.74% of drivers will be in class 1 after the system has been run for a long time (Table 6).

Research Project 2 A BMS is designed to decrease the premium of good drivers and increase the premium of bad drivers. Elasticity of a BMS is used to measure the efficiency the BMS. Mathematically, with claim frequency λ and mean premium b, the elasticity of a BMS is defined as

$$\eta(\lambda) = \frac{\lambda}{b}\frac{db}{d\lambda}$$

$$= \frac{\lambda}{b}\sum_{i=1}\frac{da_i}{d\lambda}b_i,$$

where a_i is the stationary distribution of class i in a BMS, and b_i is the individual premium for class i in a BMS. As stated above, the elasticity of a BMS measures the response to a change in the expected claim frequency and is used to compare several bonus-malus systems. An ideal BMS has an elasticity equal to 1 which states that the means of the system is perfectly balanced. Compute and interpret the elasticity of Brazilian BMS and Hong Kongese BMS as defined in Tables 1 and 3 with $\lambda = 0.05$.

Appendix

$$a_1 = \frac{p_0^6}{(1 - (p_0^5 p_5 - 2p_0^4 p_1 p_3 - p_0^4 p_2^2 + 2p_0^4 p_4 + p_0^3 p_1^3 - 6p_0^3 p_1 p_2 + 3p_0^3 p_3 - 6p_0^2 p_1^2 + 4p_0^2 p_2 + 5p_0 p_1))}$$

$$a_2 = \frac{(p_0^6 - p_0^5)}{(p_0^5 p_5 - 2p_0^4 p_1 p_3 - p_0^4 p_2^2 + 2p_0^4 p_4 + p_0^3 p_1^3 - 6p_0^3 p_1 p_2 + 3p_0^3 p_3 - 6p_0^2 p_1^2 + 4p_0^2 p_2 + 5p_0 p_1 - 1)}$$

$$a_3 = \frac{(p_0^5 p_1 + p_0^5 - p_0^4)}{(p_0^5 p_5 - 2p_0^4 p_1 p_3 - p_0^4 p_2^2 + 2p_0^4 p_4 + p_0^3 p_1^3 - 6p_0^3 p_1 p_2 + 3p_0^3 p_3 - 6p_0^2 p_1^2 + 4p_0^2 p_2 + 5p_0 p_1 - 1)}$$

$$a_4 = \frac{(-p_0^5 p_1 + p_0^5 p_2 + 2p_0^4 p_1 + p_0^4 - p_0^3)}{(p_0^5 p_5 - 2p_0^4 p_1 p_3 - p_0^4 p_2^2 + 2p_0^4 p_4 + p_0^3 p_1^3 - 6p_0^3 p_1 p_2 + 3p_0^3 p_3 - 6p_0^2 p_1^2 + 4p_0^2 p_2 + 5p_0 p_1 - 1)}$$

$$a_5 = \frac{(-p_0^5 p_2 + p_0^5 p_3 - p_0^4 p_1^2 - 2p_0^4 p_1 + 2p_0^4 p_2 + 3p_0^3 p_1 + p_0^3 - p_0^2)}{(p_0^5 p_5 - 2p_0^4 p_1 p_3 - p_0^4 p_2^2 + 2p_0^4 p_4 + p_0^3 p_1^3 - 6p_0^3 p_1 p_2 + 3p_0^3 p_3 - 6p_0^2 p_1^2 + 4p_0^2 p_2 + 5p_0 p_1 - 1)}$$

$$a_6 = \frac{(-p_0^5 p_3 + p_0^5 p_4 + p_0^4 p_1^2 - 2p_0^4 p_1 p_2 - 2p_0^4 p_2 + 2p_0^4 p_3 - 3p_0^3 p_1^2 - 3p_0^3 p_1 + 3p_0^3 p_2 + 4p_0^2 p_1 + p_0^2 - p_0)}{(p_0^5 p_5 - 2p_0^4 p_1 p_3 - p_0^4 p_2^2 + 2p_0^4 p_4 + p_0^3 p_1^3 - 6p_0^3 p_1 p_2 + 3p_0^3 p_3 - 6p_0^2 p_1^2 + 4p_0^2 p_2 + 5p_0 p_1 - 1)}$$

$$a_7 = \frac{(-p_0^5 p_4 + p_0^5 p_5 + 2p_0^4 p_1 p_2 - 2p_0^4 p_1 p_3 - p_0^4 p_2^2 - 2p_0^4 p_4 + p_0^4 p_1^3 + 3p_0^3 p_1^2 - 6p_0^3 p_1 p_2 - 3p_0^3 p_2 + 3p_0^3 p_3 - 6p_0^2 p_1^2 - 4p_0^2 p_1 + 4p_0^2 p_2 + 5p_0 p_1 + p_0 - 1)}{(p_0^5 p_5 - 2p_0^4 p_1 p_3 - p_0^4 p_2^2 + 2p_0^4 p_4 + p_0^3 p_1^3 - 6p_0^3 p_1 p_2 + 3p_0^3 p_3 - 6p_0^2 p_1^2 + 4p_0^2 p_2 + 5p_0 p_1 - 1)}$$

References

1. Columbia. (n.d.) Discrete time Markov-chains. http://www.columbia.edu/~ww2040/4701Sum07/4701-06-Notes-2.pdf.
2. Das, K. *A Stochastic Approach to the Bonus-Malus System*, Neural, Parallel and Scientific Computation, 18(3), 283-290, 2010.
3. Das, K., Sarker, S. and Diawara, N. *Further review of Panjer's recursion for evaluation of compound negative binomial distribution using R*, Missouri Journal of Mathematical Sciences, 23 (2), 182-191, 2011.
4. Dodu, D.*Comparative analysis of bonus malus systems in Italy and central and eastern Europe*, https://www.milliman.com/en/insight/comparative-analysis-of-bonus-malus-systems-in-italy-and-central-and-eastern-europe, 2017
5. Duplan, N., Hall, C., Nguyen, J., Willis, C. and Das, K. *The Mathematical Approach to Evaluating the Elasticity of a Bonus-Malus System*, The Pi Mu Epsilon Journal, 13(5), 269-276, 2011.
6. Lemaire,J. *Bonus-Malus Systems in Automobile Insurance*, Boston: Kluwer, 1995.
7. Lemaire,J. *Bonus-Malus Systems: The European and Asian Approach to Merit-Rating*, North American Actuarial Journal, 2, 26-38, 1998.
8. Edwards, A. and Das, K. *Using statistical approaches to model natural disasters*, American Journal of Undergraduate research, 13 (2), 87-104, 2016.
9. Loimaranta,K. *Some Asymptotic Properties of Bonus Systems*, Astin Bulletin, 6, 233-245, 1972.
10. Mayuzumi, T. *A study of the Bonus-Malus system*, Astin Colloquium International Actuarial Association, Brussels, Belgium, 1990.
11. Meyer, U. *Third party motor insurance in Europe. Comparative Study of the Economical-Statistical Situation*, University of Bamberg: Bamberg, Germany, 2000.
12. Okura, M.,Yoshizawa, T., and Sakaki,M. *An evaluation of the new Japanese Bonus-Malus System with no-claim and claimed subclasses*, Asia-Pacific Journal of Risk and Insurance, 2020:20190004; DOI:10.1515/apjri-2019-0004.
13. Szymańska, A. *Functions of the bonus-malus system in the motor third party liability insurance of motor vehicle owners*, Economic and Environmental Studies (E&ES), ISSN 2081-8319, Opole University, Faculty of Economics, Opole, Vol. 18, Iss. 2, pp. 925-942, 2018, https://doi.org/10.25167/ees.2018.46.28